MODERN QUATERNARY RESEARCH IN SOUTHEAST ASIA

H.R. van Heekeren 1902-1974

MODERN QUATERNARY
RESEARCH IN SOUTHEAST ASIA

papers read at the

SYMPOSIUM ON MODERN QUATERNARY RESEARCH IN INDONESIA
GRONINGEN / 16 MAY 1974

in memory of

DR H.R. VAN HEEKEREN

edited by

GERT-JAN BARTSTRA & WILLEM ARNOLD CASPARIE

A.A.BALKEMA / ROTTERDAM / 1975

CONTENTS

On 10 September 1974 Dr H.R. van Heekeren, "Oom Bob" to his friends,
died. A fascinating man has gone.

The "Symposium on Modern Quaternary Research in Indonesia", held
in the main building of the University of Groningen on 16 May 1974, was
his last symposium. And the brief excursion connected with it was his last
field work. That is why we deem it right to present this collection of symposium papers as a small memorial volume, and to reproduce some
photographs of Dr van Heekeren at the symposium on the next few pages.

We have had the pleasure of knowing Dr van Heekeren in his last few
years, but one of us (Drs G.J. Bartstra) had the privilege of co-operating
with him in fieldwork on Java and Sulawesi. Now that we re-read the books
he wrote about his expeditions, books such as "Voetsporen naar de kim"
and "De onderste steen boven", we clearly realize how fascinating and rich
Dr van Heekeren's life has been, and how many friends he must have had.
His death leaves a yawning gap.

Unfortunately, Dr van Heekeren was unable to rewrite the lecture he
gave at the symposium on the "Chronology and Archaeology in Indonesia"
for this booklet. Consequently we would like to confine ourselves to the
reproduction of the schematic outline of Indonesian prehistory that Dr van
Heekeren drew up especially for this symposium.

A number of researchers of several countries had gathered at the symposium in Groningen. Most of them knew each other, often personally, but
at any rate by name as the authors of scientific publications.

Why was this symposium held?

The Palynologische Kring of the Koninklijk Nederlands Geologisch Mijn-
bouwkundig Genootschap is the organisation of Dutch palynologists. The
Kring organizes, for its members and other persons who are interested,
meetings with lectures on a certain subject or about research in a certain
area. An "Indonesia-day" had been on the program of the Palynologische
Kring for quite some time. The fact that Drs R.P. Soejono, head of the
Bidang Prasejarah of the Lembaga Purbakala dan Peninggalan Nasional in
Indonesia, was in Holland at the time, seemed to us to be the occasion to
really organize this Indonesia-day.

In the Biologisch-Archaeologisch Instituut of the University of Groningen
researchers are now working along one of the lines indicated by Dr van
Heekeren. One of us (Drs Bartstra), in co-operation with members of the
Lembaga Purbakala, is conducting a research into the Palaeolithic Patjitan
Culture on Java. Besides, in the framework of the National Project of
Palaeoanthropological Research in Indonesia, Prof.Dr W. van Zeist has
begun palynological research on sediments that can be connected with the
earliest hominids in Java. The presence of Drs Soejono in Holland seemed
reason enough for the Biologisch-Archaeologisch Instituut also to organize
an "Indonesia-day".

The co-operation between Kring and Instituut resulted in the abovementioned symposium, the organisation of which lay in our hands.

The offer of the Firm, A.A. Balkema, Publishers at Rotterdam, to pub-
lish the lectures of the symposium in book form was highly appreciated by
both the authors and us. Part of the money necessary for the publication
of this booklet came from the Biologisch-Archaeologisch Instituut.

The articles by Prof.Dr H. Th. Verstappen and Dr D.A. Hooijer are
the extensive versions of their symposium lectures. The article by Mr T.
Harrisson is the result of a discussion that took place in a small circle of
archaeologists immediately after the symposium.

The palynological article by Mr J. Muller deals not only with the Quaternary, but also with the Tertiary. The complete paper on which the lecture
by Mr Muller and Dr J.A.R. Anderson is based will appear in the Review
of Palynology and Palaeobotany (publ.Elsevier) in 1975, and will be en-
titled: "Palynological study of a Holocene peat and a Miocene coal deposit
from N.W. Borneo". Mr Muller was so kind as to rewrite this paper for
us, and it forms, as it were, the coping-stone of the historical development of palynological research as delineated by Mrs Dr B. Polak.

The illustrations to these articles were provided by the authors themselves; Mr Jac.Klein of the Biologisch-Archaeologisch Instituut, however,
made them suitable for publication and took the photographs at the symposium.

The symposium at Groningen was no incidental event. It was, in fact,
the expression of the co-operation, recently begun, between researchers
of various disciplines in a region that raises high scientific hopes. A mul-
tidisciplinary approach towards Quaternary Studies was for a long time
the ideal of Dr van Heekeren. Drs Soejono plays an important part in this
co-operation now. The editors are happy to have been enabled to compile
the articles for this booklet as a token of that co-operation.

Biologisch-Archaeologisch Instituut
Groningen, Netherlands
Palynologische Kring of the
Koninklijk Nederlands Geologisch Mijnbouwkundig Genootschap
Leiden, Netherlands

Drs Gert-Jan Bartstra
Dr Willem Arnold Casparie

R.P. Soejono; Introduction

H.Th. Verstappen (left); H.R. van Heekeren; Mrs B. Polak; R.P. Soejono

H.R. van Heekeren (left); Mrs C. Harrisson; T. Harrisson

D. A. Hooijer (left); Mrs C. Harrisson; H.R. van Heekeren

I.C. Glover (left); H.R. van Heekeren; G.J. Bartstra

SYMPOSIUM ON MODERN QUATERNARY RESEARCH IN INDONESIA

GRONINGEN / 16 MAY 1974

R. P. SOEJONO
Lembaga Purbakala dan Peninggalan Nasional
Jakarta

INTRODUCTION

I would like to convey our thanks to those on whose initiative this symposium was organized, namely the Palynological Group of the Royal Dutch Geological Mining Society and the Institute for Biological Archaeology of the State University at Groningen, and also to take this opportunity to thank Professor van Zeist for his opening address to all the participants present.

Some of you may know that a similar symposium has been held in Yogyakarta (Indonesia) in 1972, during the ten years anniversary of the National Project of Palaeoanthropological Research. It is our feeling that the advantage of holding symposia such as this, is that we can gain more insight into the progress made in the study of mankind — from his beginnings — in Indonesia. We can also gain more understanding of our aims in this field, and increase the general interest.

Research into subjects of Pleistocene Study was started four decades ago in Indonesia. Although a great deal of individual study has been done, more expansion on collaborated work is vital to the overall results. Without inhibiting individual interest we must start to carry out systematic Quaternary research in Indonesia in such a way that a blending of the various individual studies can be achieved. Within any plan of co-ordinated work, it is always possible to observe an imbalance in the various interests and approaches. Since studies on the Pleistocene of Indonesia developed, much stress lay in the fields of palaeontology and palaeoanthropology, whereas other fields of study, in particular those connected with problems of Pleistocene environment — except for the geology of important areas of fossil men and mammals which was rather profoundly settled — were in a position of being improperly involved. Basing on a plan of co-ordinative attempts towards researching early man's life and his adaptation to the local environment, the National Project of Palaeoanthropological Research has been established in 1962. There is a need to stimulate and to extend interest to Indonesian scholars and to encourage them to improve upon their researches. For the sake of a general and overall advance in our work, these gaps in the broad outline of study should be eliminated, however difficult it may be.

Today I am able to welcome many friends who have worked in Southeast
Asia, particularly in Indonesia, and who have knowledge about the problems of these areas. Professor Verstappen who has taken an important part
in the study of geology, will give an emphasis on the development of landform and climate; Dr Hooijer will talk about the important results of his
research on Quaternary mammals; Dr van Heekeren will tell us about the
foundation and development of the human culture; Dr Polak, Mr Muller
and Dr Anderson will discuss about peat deposits in Indonesia and neigh-
bouring areas.

Several fields of Quaternary study in Indonesia have undergone thorough
research, among which the study of the physical development of man and
animal have produced increased data which open up new perspectives of
interpretation. In spite of many significant results, we are aware of the
shortcomings in the accomplishment of research in other related aspects
in the life of prehistoric man in Indonesia. The study of man's environment
in tropical areas, to which he must adapt and develop his way of life, has
been rather neglected so far, and should therefore receive more serious
attention and more systematic research.

Our thanks must be extended to the Dutch scholars and institutions for
their efforts to help to improve and increase the amount of research being
done. We would like to thank them also for their collaboration in the planning of researches which will give us the opportunity to create some more
experts of our own. Last, but certainly not least, I would like to express
our appreciation to the several American scholars, universities and foun-
dations, for their co-operation with the National Project of Palaeoanthropological Research. This co-operation has, among others, given us important results in the dating of various Pleistocene beds, from which it has
been possible to verify the age and chronology of fossil finds in Indonesia.
By combining and collaborating in our individual investigations in various
ways, and also with the aid of symposia like this, we should be able to
progress and overcome the problems involved in the study of the Quaternary.

H.Th. VERSTAPPEN
International Institute for Aerial Survey and Earth Sciences (ITC)
Enschede

ON PALAEO CLIMATES AND LANDFORM DEVELOPMENT IN MALESIA

1 INTRODUCTION

Climatic changes during the Quaternary are a subject of study of long
standing in the at present moderate humid zone where ample evidence for
alternating glacial and interglacial periods exist. More recently convincing proof has been found in the nowadays arid and semi arid belt at lower
latitudes that periods of pronounced aridity here alternated with periods
of more humid conditions. Data pointing to climatic change in the humid
tropics during the Quaternary remained few in number, however, particularly with respect to South East Asia. The evidence available for this
part of the world and especially for Indonesia, will be reviewed in this
article and evaluated in the framework of the general atmospheric circulation patterns of the past and the present. Changes in temperature and,
more important, in rainfall and humidity will be considered as will the
changes in sea level, vegetative cover, soil formation and landform development. The accumulated data, to the author, are sufficiently strong conver-
ging evidence for the assumption of considerable climatic and environmen-
tal changes during the Quaternary also in this part of the world. This is in
concordance with humid tropical areas in Africa and Latin America.

The author gratefully acknowledges the critical comments on the manu-
script made by Prof.Dr F.H. Schmidt, Dr C.J.E. Schuurmans and Drs H.
J. Krijnen of the Royal Netherlands Meteorological Institute (KNMI), De
Bilt, by Prof.Dr C.G.G.J. van Steenis and Dr J. Muller of the Rijksherbarium, Leiden and by Dr J.J. Nossin, ITC, Enschede.

2 POSTULATED PALAEO CLIMATES

2.1 Generalities

Three factors have exerted a great influence on the climatic conditions of
the past in SE Asia and particularly on those that seem to have prevailed
during the Pleistocene glacial periods.

A first, important, factor is the position of the Intertropical Convergence zone (ITC) which is affected by the development of the anticyclones over Asia and Australia and thus fluctuates with the season but also with secular changes in atmospheric circulation. Since most of the precipitation in the area is associated with the ITC, marked changes in the distributional pattern of the rainfall must have occurred when, during the glacials, the anticyclone over Asia, (and also, in all likelihood, over Australia) was more strongly developed than at present.

A second factor is the world wide drop in air and sea water temperature, during the Quaternary ice ages which caused a lowering of the snowline and the forest line and affected the altitudinal zonation of vegetation in the area.

A third factor is the emergence of both the Sunda and Sahul shelves during the glacials due to the lowering in sea level. This may have caused increased dryness particularly in lowland areas. Each of these three factors influenced a whole range of meteorological and climatological elements such as precipitation, evaporation, humidity, wind-direction, cloudiness, sun shine, temperature, etc. Changes in precipitation/humidity on the one hand and in temperature on the other are of particular interest in the study of palaeoclimates. It seems that the position of the ITC mainly affected the former and that the drop in air and seawater temperatures also influenced the precipitation and humidity conditions. The greater extension of the land areas in the glacial periods had an effect, of subordinate magnitude (?) on both precipitation/humidity and temperature. The three factors will be discussed separately below.

2.2 Changes in the position of the ITC

The Intertropical Convergence is the zone where the rising part of the Hadley circulations of the northern and southern hemispheres meet. The 100.000 times broader descending parts of these circulations are found in the southern and northern mid-latitudinal dry belts. The trade winds represent the lower horizontal connection and have their counterpart in the tropopauze where the air that rose in the ITC moves poleward. This circulational system affectuates the evacuation of latent heat out of the tropical zone that before has been transmitted to the tradewinds blowing over the tropical seas.

The position of the ITC is not exactly over the equator, but is affected by the distribution of land and sea and it also changes with the season. These changes have a particularly large latitudinal range where large landmasses occur such as in South and SE Asia and Africa and are minimum over the mid-Pacific Ocean where it only amounts to a few degrees latitude. The average position of the ITC in SE Asia during the northern winter and summer respectively is given in Fig.1. The position in December-February is roughly parallel to the equator at about $10-12^{\circ}$ S and it thus passes south of Java and over the northernmost part of Australia. In July-September, however, the ITC is much more irregular in outline and has much more northward position. It is situated at about 23° N near the Himalayas and the Gulf of Bengal, reaches around 32° N at the south coast of

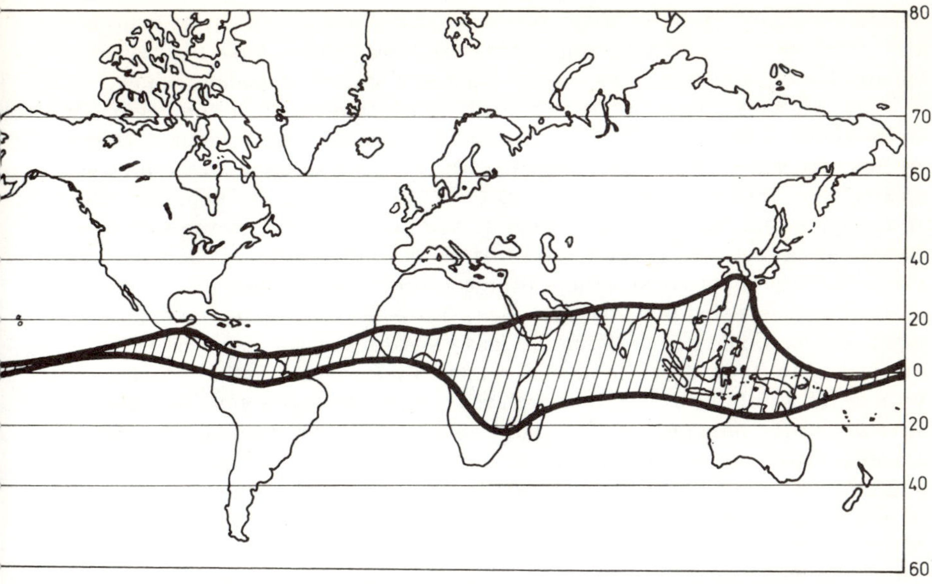

Figure 1. Average annual latitudinal shift (hachures) of the Inter Tropical Convergence zone (ITC). The southernmost thick line indicates the average January position and the northernmost thick line the July position of same. The hachured zone is particularly broad in SE Asia and thus the climate of much larger areas will be affected there if deviations of the (present) average position of the ITC occur than in the Pacific and some other oceanic areas where the annual shift is much less pronounced.

Hunshu, Japan, and further to the east rapidly takes a more southward position again to reach the equator approximately to the east of New Guinea.

The above mentioned approximate seasonal extremes in the position of the ITC are, of course, only indicative for the average conditions at present prevailing, and fluctuations can be expected over the years, depending on the development of the Asiatic anticyclone in the first place and the Australian anticyclone in the second. An above-normal development of the Asiatic anticyclone will cause a more southerly position of the ITC in December-February and will retard and reduce its subsequent northward displacement during the northern summer. The above-normal development of the smaller Australian anticyclone will have a comparable, though probably more limited effect and cause a more northerly position of the ITC in July-September and a retarded and reduced southward shift in the southern summer. It stands to reason that the effect of the fluctuations of the Asiatic anticyclone dominates the larger part of the area whereas the simultaneous fluctuations of the Australian anticyclone will be felt to some extent in the eastern Lesser Sunda Islands, the Moluccas and New Guinea.

The simultaneity of the ice ages in the northern and the southern hemisphere suggests that also during the Pleistocene the fluctuations of the Asiatic and Australian anticyclones were in phase. The ITC certainly had

a more southerly position than at present in the whole area or at least in
the major part of it. The climatic changes provoked by these fluctuations
in position of the ITC are most important and will be discussed below.

Climatic variations that have occurred in Indonesia since the beginning
of the meteorological observations about one century ago have been studied
by Schmidt and Schmidt-Ten Hoopen (1951). The variations proved to be
strongly correlated to the position of the ITC and to the development of the
Asiatic and Australian anticyclones. It was demonstrated on the strength
of wind data that the ITC in some years moves far south of Java and
remains there for a considerable time whereas in other years it assumes
a stationary position over the Java Sea. Since the maximum amount of
precipitation occurs in the form of heavy showers at and near the ITC,
considerable fluctuations in annual rainfall thus result in all stations. Further to the south the rainfall, though more continuous, is considerably
less and to the north of the ITC is of a much smaller magnitude altogether.
Consequently, heavy showers result in high precipitation figures over Java
and other parts of southern Indonesia if the ITC moves far south, whereas
high precipitation values are recorded over northern Indonesia and lower
values in the south when it remains further north, e.g. over the Java Sea.
These fluctuations are considerable and, for Pontianak for example, amount
to 25% of the annual mean if 11-year periods are averaged and even to
75% if the highest and lowest annual values on record are considered.

It is noteworthy to emphasise that the northern stations and the southern
ones at present respond in opposite ways to fluctuations of the Asiatic
anticyclonal development and the corresponding position of the ITC (Fig. 2).

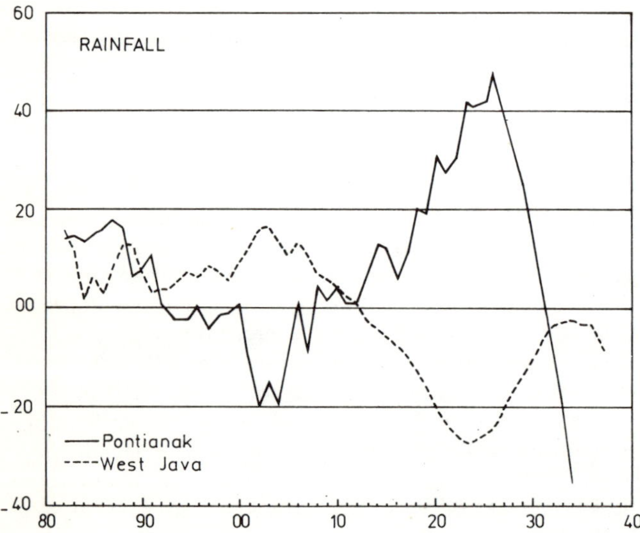

Figure 2. Fluctuations of the 11-year totals of the yearly amount of rain at Pontianak and in West Java in the period 1882–1938. (Schmidt en Schmidt-ten Hoopen,
1951). The opposite trend in northern and southern Indonesia is evident.

6

This is so because the ITC in any case remains in or near the area. It
stands to reason, however, that when during the glacial periods the ITC
was able to move further south, the stations in southern Indonesia also
had low precipitating values thus following the trend that now still characterizes northern Indonesia. A greater length and intensity of the dry season
tends to be associated with a decrease in precipitation and parts of the area
falling within the Af climate, in some years have Aw characteristics. Fig. 2
illustrates the opposite trend in precipitation changes in northern and
southern Indonesia using Pontianak and Jakarta as examples and Fig. 3
gives the distribution of the extremes of rainfall in Indonesia in 1900 and
1925 respectively, after Schmidt and Schmidt-ten Hoopen (1951). It
should be noted that the thick line separates the stations of above-normal
precipitation from those having below-normal deviations. It does not

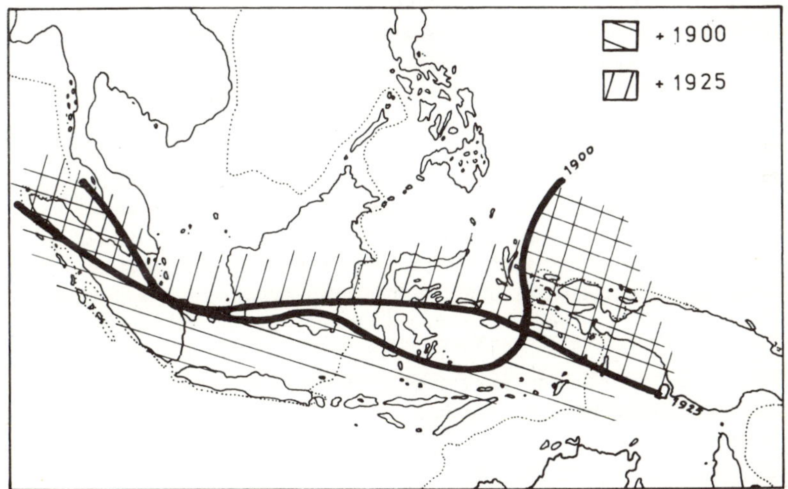

Figure 3. Distribution of the above normal rainfall about 1900 and about 1925
(Schmidt en Schmidt-ten Hoopen, 1951). The areal extent of the opposite trend
indicated by Fig. 2 is clearly demonstrated.

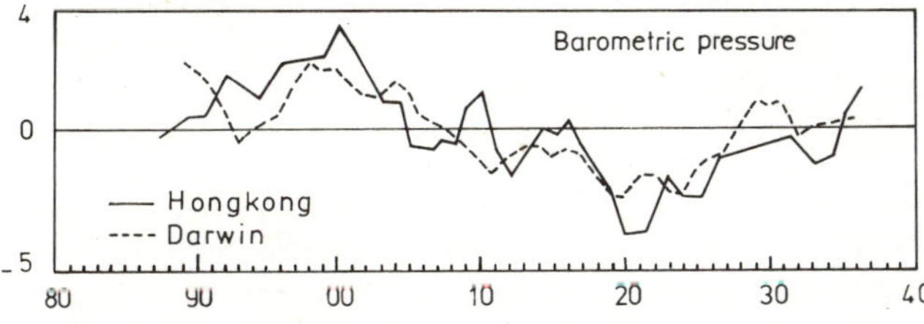

Figure 4. Fluctuations of the 11-years totals of the yearly pressure at Hongkong and
Darwin in approximately the same period as covered by Fig. 2 (Schmidt en Schmidt-
ten Hoopen, 1951). Note that the fluctuations of the Asiatic and the Australian Highs
are in phase and through their effect on the position of the ITC cause the fluctuations
in rainfall of Fig. 2 and 3.

represent the position of the ITC which is situated further to the south. Comparison with Fig. 4, showing the 11-year totals of the annual precipitation at Hongkong and Darwin which reflect the fluctuations of the Asiatic and Australian anticyclones respectively, informs that around 1900 the anticyclones were strongly, and around 1925, weakly developed.

2.3   Changes in temperature

The drop in temperature that has occurred in the equatorial regions during the Pleistocene glacials has been underestimated until recently. Büdel (1957) still states that in the tropical rainforest it amounts only to about $1-2^0$ C at groundlevel and to $3-4^0$ C in the upper troposphere. It was thus considered of substantially smaller magnitude than in the moderate climate where e.g. a drop of $10-14^0$ C at groundlevel was assumed for the central European lowlands. Two observations suggest that the Quaternary temperature changes in the humid tropics, were somewhat larger, however.

Firstly there is the fact that a rise in average annual temperature of about 1^0 C has been recorded at most equatorial meteo stations in the last 100 years. If the minor climatic fluctuations of the recent past had already such a pronounced effect on the temperatures in this climatic zone it stands to reason that considerably greater temperature changes were associated with the glacial-interglacial alternations that characterised the Pleistocene period. However, part of the recent rise in temperatures appears to be due to increased urbanization at or near the meteo stations concerned and this fact, of course, reduces the validity of this argumentation.

Secondly it is difficult to account for the sinking of the snow line of approximately 1000 meters in the Central Mountains of New Guinea (Dozy 1937; Verstappen, 1952, 1960, 1964; Bik 1967; Reiner 1960; Löffler 1970, 1971), if indeed the lowering in temperature amounted only to $1-2^0$ C at ground-level. The observations on a Pleistocene glaciation in the top area of Mt Kinabalu, Borneo, with a snowline at 3600-3700 meters m.s.l. (Koopmans and Stauffer, 1968; Stauffer, 1968) pose the same problem and are in concordance with comparable observations in equatorial alpine mountains elsewhere in the world. If one adheres to the view that the Central Mountains of New Guinea have continued to rise after the Pleistocene glacials the problem becomes even greater although one has to consider that the position of the snowline is a complex matter and is affected not only by temperature but by precipitation, humidity, etc as well.

Evidence of a greater amplitude of the Pleistocene temperature changes has since been obtained. Global estimates indicate a decrease in mean temperature of $3-9^0$ C (depending upon latitude, etc) during the last glaciation (Flint and Brandtner, 1961; Fairbridge, 1967). Kraus (1973) mentions even larger changes, and on the strength of the newest evidence maintains that the seawater temperatures in the northern Atlantic Ocean (above 50^0 N) dropped 10^0 C and that the waters of the equatorial Atlantic Ocean and the Caribbean were $3-4^0$ C cooler than today. Over the continen-

tal lowlands the differences were even greater; the summer temperatures
over the formerly glaciated parts of Europe and north America are estimated to have been about 30^0 C lower than to date. Galloway (1973) maintains that the ablation season in Australia and New Guinea was substantially colder than 6^0 C below present values.

Investigations of deep sea sediment cores (Ericson and Wollin, 1956;
Ericson, 1961; Emiliani, 1955, 1966; Emiliani and Milliman, 1966; Oba,
1957; Parker, 1962, 1967) have played an important role in these new
views. Fairbridge is of the opinion that sea water temperatures were
depressed by about 5^0 C during the glacials. The data available seem to
justify a drop in average temperature of $3-5^0$ C in equatorial regions during the ice ages and possibly a rise of $2-3^0$ C with respect to present
values during the interglacials.

Even a lowering in temperature of $3-4^0$ C or even 5^0 C at sea level possibly cannot fully account, at the present lapse rate of approximately
0.6^0 C/100 m, for the lowering of the snowline of 1000 meters observed in
SE Asia. An increased lapse rate thus seems to have prevailed during
glacial conditions. Kraus (1973) counts in the upper troposphere with a
three-fold amplification of the changes in surface temperature of the seawater under equatorial circumstances due to a decrease in condensation
heat. The climate in the mountains consequently would become considerably cooler as a result of even minor depressions of sea water temperature.

2.4   Changes of sea level

The world wide glacio-eustatic variations in sea level during and after the
Pleistocene are a third way in which the ice ages made themselves felt in
the SE-Asian region. There is some discussion as to the exact amplitude
of these changes but it is generally assumed that during the last glaciation
the sealevel was about 100 meters lower than to date. Pfannenstiehl (1953/
54) mentions 91 meters whereas some other estimates even are in the
order of 150 meters. The different estimates are due to the posterior
crustal movements that have occurred in many places.

Offshore exploration for oil and minerals in the Sunda shelf in recent
years has added considerable new evidence on Pleistocene sea levels in
SE Asia (Biswas, 1973; Aleva et al., 1973). Biswas mentions withdrawals
of 60-70 meters (205-236 ft) for the youngest regressional phase which
occurred at the Pleistocene-Holocene interface and was radio-carbon
dated at 11,170 $\pm$ 150 years BP. Two, possibly three, older Pleistocene
periods of low sea level were also recorded and were in the order of 45-
70 meters (150-200 ft). These figures are rather conservative as compared to the values obtained in other parts of the world. This discrepancy
may point to a slight postglacial rise of the Malay peninsula and its surroundings. Alternatively it may be that further on seaward of the sample
sites studied by Biswas the same sequence of deposition and vegetation
can be observed at some greater depth and has occurred at a somewhat
earlier time when the sealevel was still lower.

Whatever the exact amount of glacial depressions of sea level may have
been, it is evident that due to the great extension of shallow shelf areas
(Sunda and Sahul shelfs) in SE Asia the land surface in glacial times was
much larger than at present. More than $3,000,000 \text{ km}^2$ of shallow warm
seas were converted into land. This fact, and the somewhat lower temperatures, may have caused a reduction of the evaporation particularly since
large tracts of land areas in those days were probably not covered by tropical rain forest due to the decrease in precipitation caused by the shift in
the position of the ITC. The larger diurnal changes in temperature that
probably occurred will have strengthened the valley winds at places, thus
adding to the relative dryness of the lowlands and bringing some rain in
the higher parts.

Another aspect worth mentioning is the effect of changes in ocean currents. In the east the waters of the south equatorial current could not reach
the area since the Torresstrait ran dry. In the west, the seasonal currents
that at present transport relatively cool water to the equator between the
Asian mainland and Borneo and the Philippines during the northern winter
and that evacuate warm water from the Sunda shelf-area to the north during the remainder of the year, became ineffective since the South China
Sea for the major part ran dry. Seasonal changes in mean daily temperatures thus increased. The change in ocean currents thus added to the con-
tinentality caused by the increase in landarea and the resulting decreased
influence of the thermally "inert" ocean waters.

2.5   Postulated Pleistocene climatic conditions in SE Asia

It will be evident from the aforesaid that the glacial periods, also in SE
Asia, were characterized by lower precipitation and temperature values
than prevail at present. The most recent occurrence dates back to 11.170
years BP and two or three pre-Würm occurrences also have been proven
(Biswas, 1973). Flohn (1952) estimates that due to lower temperatures
the global evaporation and consequently the precipitation was at least 20%
less than to date. Galloway (1965) believes that a drop of 30% or more
is improbable and maintains that also in middle and low latitudes the
precipitation was markedly lower than to date. Schmidt and Schmidt-ten-
Hoopen (1951) pointed out that the comparatively small secular fluctuations
in the mean position of the ITC that have occurred in this century have
caused a decrease in annual precipitation of 25% (e.g. Pontianak). It stands
to reason that the larger fluctuations of the ITC during the Pleistocene
have had at least the same effect. It thus seems not too hazardous to
assume a drop in annual rainfall in the order of 30% below present values
during the Pleistocene glacial periods. A factor of particular importance
is the more pronounced dry season then characterizing the climate. It
provoked drought stress in the vegetation and caused changes in the geomorphological processes. The reduction in rainfall in Malesia is believed by

The reduction in rainfall in Malesia is believed by the author to have
been equal to, if not somewhat more pronounced than, that in most other

10

equatorial areas due to the combined effect of the global drop in temperature, the shift in position of the ITC and the greater extension of land areas. In parts of the area now having an Af climate an Aw climate prevailed instead. Since the area is at present a comparatively humid part of the equatorial belt it was nevertheless during the glacials probably less dry than most parts of equatorial Africa and Latin America. It should also be understood that higher precipitation values occurred at greater altitudes and possibly also near the southern margin of the Sunda shelf where the marine influence was stronger and the ITC nearest.

The seawater temperatures were about $4-5^0$C lower than today. An even somewhat greater drop may have characterized the average temperatures of the lowlands where the diurnal temperature changes were increased due to the greater dryness. The snowline was about 1000 meters depressed and the climatological zonation in the mountains was thus vertically compressed. During the interglacial periods and also during the Mid Holocene climatic optimum and in the Upper Tertiary the climate was probably more humid and about $1-2^0$C warmer than at present.

3 FLORAL AND FAUNAL RESPONSE

3.1 Vegetative patterns and climatic change

Botanical studies several decades ago already revealed that the distribution of species characteristic for the equatorial rain forest and even more so the distribution of drought plants associated to tropical climates with a pronounced dry season, cannot be explained if the climatic pattern at present prevailing in SE Asia has always characterized the area also in the past. Van Steenis (1935a; 1939) was the first to point out the increasingly dry and seasonal climate between Burma and Java when during low, glacial, sea levels the land area in the region was considerably larger than to date. A small change in climate, according to him, would result in a major shift of vegetation boundaries; even now, in dry years, monsoon-forest conditions are found in the humid parts of Sumatra and Borneo.

Nevertheless, botanical evidence suggests that drier conditions have been rather the exception in the past and humid tropical conditions are thought to be the more "normal" situation at least in insular SE Asia. Burkill and Holttum (1923) are of the opinion that humid tropical conditions occurred uninterruptedly from East Africa to the Pacific during the Miocene. The large rain forest area only became disrupted when later drier conditions set in whereafter the three rainforest cores still existing, developed separately. Van Steenis in his publications (1935b; 1947, 1961, 1964, 1965) emphasizes that the Dipterocarpaceae, typical representatives of the equatorial rainforest, have occurred uninterruptedly in Sumatra and Borneo since at least the Miocene and thus also adheres to the view that cores of tropical rainforest have persisted throughout.

The distribution of drought plants in the area is of particular interest, as they all show a disjunction between lower Burma and eastern Java.

This applies e.g. to teak (Tectona grandis L.f), sandalwood (Santalum album L.) trees, to many genera of the Papilionaceae and to several Leguminoseae and grasses. In total hundreds of plants thus occur both in the western and the eastern parts of SE Asia and are separated by the core of humid tropical rain forest that covers the western part of the Malay archipelago. Part of these are of Asiatic origin, others of Australian type or both Asiatic and Australian type. East and westward migration thus seems to have occurred. Only very few drought plants are endemic, however, which points to the fact that there never have been long-lasting periods of drier climate. The exchange of drought plants must have been a geologically rather recent occurrence since practically all species are the same as those found in the Australian and Asiatic continents. It thus stands to reason to assume, with Van Steenis, c.s. that the migration occurred when during the Pleistocene glacial periods a tropical climate with pronounced dry season was more widespread than at present. Van Steenis maintains that the migration was mainly bound to a "corridor" or pathway connecting continental Asia with Australia via the Philippines, Celebes and Moluccas, Java, Lesser Sunda Islands and southern New Guinea (Fig.5) although he considered the possibility that Sumatra also may have been the scene of migrating vegetation. One may wonder whether the importance of this "corridor", which is mainly substantiated by the present distributional patterns of climate and vegetation, is not somewhat overestimated by Van Steenis. As we know that the amplitude of the Pleistocene climatic variations in the humid tropics are distinctly larger than formerly assumed, one can envisage that the "corridor" was very broad

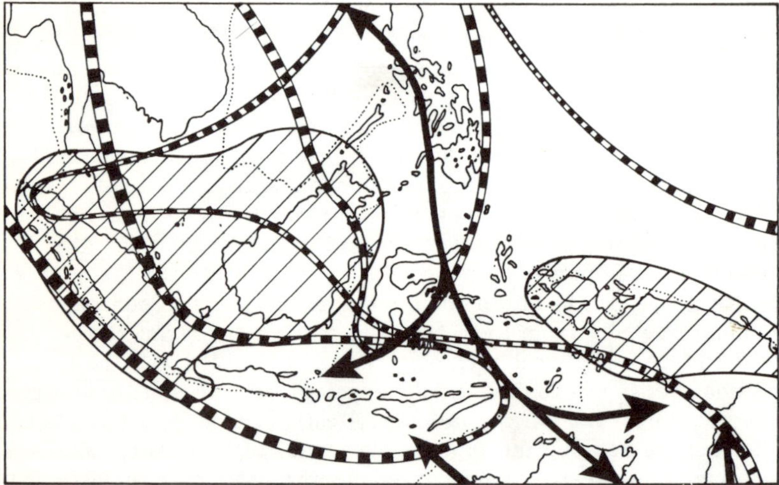

Figure 5. The two large cores of everwet rainforest on the Sunda shelf and New Guinea (hachures), the migration tracks of seasonal plants (black arrows) and the three migration tracks of cold-loving mountain plants from temperate regions into Malesia (block bands), (van Steenis, 1961).

12

and poorly defined. In effect monsoon forest and in the driest parts (tree)
savannah may well have covered the larger part of the humid tropical
cores at either side, reducing the rainforest there to hilly and other more
humid parts. Studies on Pleistocene pollen will hopefully soon solve this
question. The abovementioned reflections do not basically affect Van
Steenis' concepts but only accentuate the effects of climatic change even
stronger than he ventured to do.

In interglacial periods and in postglacial times the rain forest expanded
again and even beyond its present limits. One may assume that it shrank
in size the last few thousand years after the Mid Holocene climatic optimum.
The few wet-vegetation "islands" now occurring in some areas with marked
dry season e.g. in eastern Java (Van Steenis, 1965, p. 67) are of interest
in this connection.

A noteworthy attempt at reconstruction of the vegetative patterns of
Australia, the Sahul shelf and New Guinea, in glacial and postglacial times
is the one made by Walker (1972). Postulated changes in temperature
precipitation and vapour pressure are used to arrive at a computerized
simulation of the vegetation during the various climatic regimes that are
believed to have prevailed in the area during and after the last ice age.
Although one may wonder whether the postulated climatic conditions completely correspond with the situation that actually prevailed, it is evident
that important vegetational changes have indeed occurred. Nevertheless it
remains equally true that a core of equatorial forest remained in existence
in most of northern and central New Guinea during the dry glacial periods
and that arid vegetation and scrubland during the humid interglacials and
the postglacial optimum never completely disappeared in central Australia.

The distribution of mountain plants, so far, has not generally been considered in the context of climatic change, although the disjunctions occurring in these "cold" loving plants also are difficult to explain under present
climatic conditions. The fact that, until recently, the temperature decrease
in the humid tropics during the glacials was thought to be negligible ($1-2^0$ C)
perhaps sufficiently accounts for this. A drop of temperature in the tropical mountains of 5^0 C during Pleistocene glacials as e.g. the lowering of the
snowline seems to suggest, may have played a role in the distribution of
certain mountain plants which cannot be neglected. The migration of other
mountain plants may also partly have occurred in earlier periods, however, when a considerably higher relief existed in parts of Malesia, e.g.
the Upper Tertiary. The migration tracks of cold-loving mountain plants
are indicated in Fig. 5. Their age, at present, cannot be specified.

3.2   The Malesian fauna and climatic change

The response of the fauna to the changes in climate, vegetation and sea
level that have occurred in the area since at least the Upper Tertiary is a
means of estimating the magnitude and effect of these changes. Molengraaff
and Weber (1920) already mentioned the great similarity of the freshwater
fish fauna of western Borneo and southeastern Sumatra as evidence sup-

porting their view, that the rivers of both these areas, during the Pleistocene low sea levels, formed part of the upper reaches of the great Sunda River that drained northward in the area at present occupied by the South China Sea. The freshwater fish fauna of eastern Borneo is rather different as these rivers never formed part of the Sunda river basin. The Pleistocene sealevel changes and the repeated collation and disruption of the drainage systems so caused, thus are clearly reflected in the present distributional pattern of the freshwater fish fauna.

Fossils of large land mammals are a subject of particular interest in the context of this study. The publications on this subject are numerous (e.g. Dubois, 1890, 1937; Hooijer, 1949, 1968; v. Koenigswald, 1940, 1950) and are often related to studies on Pleistocene man. The fauna is of Asian origin in the western part, whereas in New Guinea a poor fauna of Australian origin occurs; the fauna of the islands between the two shelf areas is mixed in character.

The migration of the various types of large land mammals into the area is governed by the topographical and climatological barriers that existed in the Pleistocene. As the Sunda shelf ran dry repeatedly during the Pleistocene low sea levels it is evident that from a topographic point of view, the mammalian (land) fauna could spread eastward as far as eastern Borneo without major difficulties. The Makasar Straits (Wallace's line) have long been considered an insurmountable barrier for faunal migration. In the last 25 years, however, large Pleistocene land mammals have been found also on islands situated to the East of the Sunda shelf such as on Celebes (by H.R. van Heekeren), Flores, Timor and Mindanao. This seems to indicate that the topographical barriers between the two shelves during the Pleistocene were less formidable than they are at present. Sartono (1973) estimates that apart from a western migration route across the Sunda shelf also a northern migration route via Taiwan and the Philippines to Celebes and the Sundaland existed and that the western (not the northern) route extended also the Lesser Sunda Islands. Thus landroutes must have existed, according to Sartono, and were disrupted by major tectonic movements during the latter part of the Quaternary (see also Charlesworth, 1957). The increased volcanic activity towards the end of the Pleistocene in Java and Sumatra, among others resulting in the thick rhyolitic ash deposits observed by Stauffer in West Malaysia between Kuala Lumpur and Penang and dating from 30,000 BP (personal communication by P.S. Ashton) may well be associated with the renewed tectonic activity assumed by Sartono.

For our purpose, the effect of climatological barriers on the types and distribution of some Pleistocene mammals is of even greater importance than the role of topographical barriers. Several fossil mammals found on Java and other islands, such as antelopes, hippopotamus and the Bubalus Palaeokerabau are grazing and/or browsing animals rather adapted to more or less open vegetation. They could only have spread with difficulty under the present equatorial rainforest conditions. Fossil land mammals may thus also contribute to our understanding of the palaeo-climatological

conditions in the area. An interesting paper on the subject is by Medway
(1972) who maintains that the Lower Pleistocene mammalian fauna of Java
has affinities to the Siwalik Villafranchian but is considerably poorer
since animals adapted to scrub and savanna vegetation, like the true
horses (Equus), the camel, giraffe and several antelopes and other
bovines were prevented from migrating to Java by a forest barrier that
has existed since the Pliocene. The herbivores of Java are adapted to
riverine, forest edge or forest habitats.

During the Middle Pleistocene a large number of forest-dwelling species
invaded the western area and it is evident, that, according to Medway, on
one or more occasions during this period forest-adapted animals could
move freely over the Sundaland. However, some animals were clearly
adapted to browsing and grazing, their diet being rich in bamboo, grass,
etc. These observations may be used in favour of the hypothesis that the
Mid Pleistocene vegetation of the Sundaland was less homogeneous than
the present rain forest, and was more like the alternation of rainforest,
monsoon forest and scrub/grass now characterizing large tracts of continental SE Asia.

The world-wide extinction among mammals in the Upper Pleistocene
also affected the Malesian region, though to a somewhat lesser degree
than in other parts of the tropics. Those who failed to survive till present,
were particularly the grazing mammals, which fact finds an easy explanation in the assumption of a renewed humidity increase of the area. The
fact that the Upper Pleistocene extinction of mammal genera was less
pronounced in SE Asia than in other parts of the world may be due to the
less drastic changes in climate and vegetation as compared to the temperate zone and also to a somewhat higher rainfall as compared to most other
equatorial zones of the world (Peterson, 1970). Aridity, was never
so severe that steppe or grass savannah could establish itself in Malesia.
Monsoon forest and tree savannah were probably the characteristic vertebrate environment during the glacials (Peterson, 1970).

It is of interest in this context that also in the southeastern part of
Australia there is some evidence for a shift in emphasis, most marked
about 10,000 BP, from the drier open woodland or heath dwellers to
mammals of the wetter forest or dense thickets (personal communication
by D. Merrilees, Palaeontologist, Western Australian Museum, Perth).

Also the lower temperature in Malesia during the glacial periods,
according to some, may be reflected in the distributional patterns of
mammals. At Niah, Sarawak, the insectivore Hylomys and the ferret-
badger Melogale were found near sea level in Upper Pleistocene beds,
whereas their present occurrences are disjunct and restricted to the montane zone of Borneo. The distribution of the rat Rattus ochraceiventer on
Borneo also has been interpreted as evidence for cooler climates by Medway (1964b, 1972) who states that a temperature rise of 2-3, 5^0 C would
suffice to account for the altitudinal rise of their habitats of 300-1000
meters. Since the bones are found in human habitats the possibility that
hunters brought these animals from the hills to the caves, cannot be

excluded, however, and this evidence, in itself, therefore is not very
convincing.

4. EFFECTS ON LANDFORM DEVELOPMENT

4.1 Morpho-dynamical changes

Many geomorphological processes, generating the landform development,
have varied substantially in type and intensity with the alternating humid
tropical and drier, slightly cooler, seasonal climates that have characterized the area during the Pleistocene.

Intense chemical weathering dominated in the interglacial equatorial
rain forest conditions as it does in the Holocene. Deep weathering and soil
formation ultimately resulted in the formation of clays carried off in suspension by the rivers, the bed load of which was comparatively unimpor-
tant. Sliding, slumping and other mass movements were important slope
forming processes in many parts. The dense tropical vegetation imposed
linear river work but intense ravination was mainly restricted to Tertiary
(and more recent) beds of low resistence. Denudation rates were very
high, particularly where weak rocks outcrop in tectogene areas, such as
in the riverbasins of Java on which Mohr (1908) and L.M.R. Rutten
(1917) based their calculations.

The denudation and accretion rates have multiplied by several times,
particularly on Java since in the last one or two centuries massive deforas-
tation has occurred. There is also ample evidence of accelerated coastal
accretion in this period. Although tectonics and sea level changes have an
important influence on the horizontal displacement of the lowland coasts in
SE Asia, there is no doubt that the last two centuries geomorphologically
are to be considered as an anomaly.

The alluvial plains formed along many coasts in the last 5000 years are
mainly clayey in texture. Natural levees and back swamps are the main
relief elements. Beach ridges, marking present or former coastlines,
only occur where a source of sand and sufficiently strong onshore winds
are available for their formation.

The conditions and processes were rather different when during the
glacial periods monsoon forest and (tree) savannah prevailed in the area.
The lower precipitation, humidity and temperature values and particularly
the more pronounced seasonality of the climate brought about a marked
reduction in chemical weathering. Physical disintegration of rocks became
of more importance, but particularly in the high mountains where frost
shattering, at present restricted to a few isolated peaks in central New
Guinea rising above 4000 meters m.s.l., made itself felt from approximately 3000 meters upward. The lowering of the forest line and of the ver-
tical vegetation zones also added to the increased production of coarse
debris in the mountains.

The decrease of vegetation favoured the evacuation of the clayey waste
previously formed. The weathering mantle and soils formed during the

glacials tended to be thinner and — lithology permitting — coarser textured. Slides were a less important factor in slope development. Ravines in the hills became torrential in nature; diffuse run-off and sheetwash occurred on gently sloping terrain, processes that were prevented from developing during the interglacials by the dense forest cover that imposed linear runoff.

The first mention of soil development during drier Pleistocene periods which differs from the present is made by Mohr (1922), who describes black clayey soils from eastern Java underlying the Holocene alluvial soils and originating from weathering in an alkaline environment in areas with a particularly dry monsoon. They are also found in some borings even below present sealevel. Evidently these black soils were drowned by the postglacial rise of sea level. It stands to reason that the soils formed in the humid tropical environment of the Upper Tertiary were modified or replaced by other soil types in the drier conditions prevailing during the Pleistocene glacials. The decrease in vegetation resulted in decapitation of the ferralitic soils in the hills and deposition in the lower parts. Andosols were formed in the volcanic materials and brown and yellowish lithosols in the non-volcanic materials. Climatic change is also well-demonstrated by the soil development on the large fluvio-volcanic fan occurring near Bogor (Java) where deep-red latosols and locally laterites are underlying considerably younger brown latosols. The former point to a more rapid decomposition of organic matter and indicate a marked dry season, whereas the latter correspond better to the present climatic conditions near Bogor where the dry season is not very distinct (oral communication Dr J. van Schuylenborg, University of Amsterdam). Podzolidation of previously formed latosols observed in many localities in Malesia, may point in the same direction. Ashton (1972) states that among the deeply weathered soils that characterize the soils of the Malay peninsula about 6% of the total area is occupied by laterites and lateritic soils. They occur on relics of and old erosion surface the age estimates of which range from Early Quaternary to Cretaceous (Eyles, 1967, 1970). Bauxites are also associated to this.

The drier (glacial) climate, particularly the increased seasonality and the larger daily temperature changes, also must have had a pronounced effect on the river regime. It stands to reason that the debris formed by alteration and disintegration was transported by the high discharges of the wet monsoon. With the advent of the next dry season, however, particularly the coarser material was probably temporarily deposited in the river beds. Large tracts of the river valley were thus characterized by a valley fill that, with time, gradually increased in thickness. During the dry monsoon the discharge of the river was strongly reduced, possibly even to a trickle of water, whereas during the wet season occasionally the whole riverbed may have been filled with water. Rivers under such conditions tend to be braiding rather than meandering.

One can obtain a fairly good impression of the river regimen that must have prevailed in large parts of Malesia during the glacials, since this

kind of run-off characteristics is still found at present in the easternmost
Lesser Sunda Islands (Verstappen, 1955, p 138) and in Indonesia is known
as "Timor River". After the postglacial rise in sealevel and the transgression so caused, these glacial valley fill and fan deposits near the rivers
often border inland on the fine-textured alluvial plains with natural levees
and beach ridges formed since the Mid Holocene as already mentioned earlier. Coarse gravel fans were formed at the foot of mountain-fronts where
a sudden decrease in river gradient occurred with associated deposition of
bedload. Such fans are among others well-developed along the steep southern side of the Central Mountains of New Guinea, on the island of Ceram,
etc. At present fan building is restricted to areas where sufficient supply
of sandy or pebbly material is provided, usually by volcanic sources.

The humid tropical conditions, leading to intense chemical weathering
and following the postglacial rise in sea level, had a profound effect on
off-shore sedimentation in the shelf seas. The clays carried off in suspension by the rivers were deposited on the bottom of the shelf seas in the form
of a nearshore clay blanket. The occurrence of such deposits was first
reported off the north coast of Java by Mohr (1919), who together with
White produced a map of the sediments found in the Java Sea. The matter
was also discussed and the map reproduced by Molengraaff (1920) and by
Baartmans et al. (1947). Some new data were obtained by Neeb (1943) and
by Van Baren and Kiel (1950) who distinguish several sediment-petro-
graphical provinces. Verstappen (1953a) gives some details in the Bay of
Djakarta and comments on the conclusions arrived at by Van Baren and
Kiel (p. 53). The sediments of the Malacca Strait have been investigated
by Keller and Richards (1967). Cores sampled showed a stiff indurated
silty clay with peat representing a Late Quaternary surface C-14 dated
10,000 BP.

The Holocene clay blankets, which in recent years have also been
reported from other tropical shelf areas e.g. in South America and West
Africa have been surveyed in the sea area between Singkep and Bangka
islands by way of acoustic continuous profiling and drilling (Aleva et al,
1973). The areal extent and thickness of this younger sedimentary cover
is mapped in some detail and is described as a flat-lying grey mud formation, rich in shell fragments. These fine clastic deposits in the areas
studied are supposedly derived from Sumatra and Borneo respectively.
Punch core evidence off the east coast of Malaya also revealed the existence of a blanket of marine clays (Biswas, 1973).

It is noteworthy that underlying the blanket a mottled or spotty red clay
horizon has been reported by Aleva et al, pointing to subaerial development, desiccation and oxidation and soil formation. Locally peat has been
found which likewise indicates emergence during the warm glacial regression. Biswas mentions laterisation and lateritewash from the Holocene-
Pleistocene interface. The coarser bottom materials, such as loamy sand,
sand and gravel, according to Mohr and White occurring further to the
north in the central parts of the Java Sea and near Borneo contrary to the
clays just mentioned, contain iron concretions which fact points to a grad-

ual submergence of these older formations during the postglacial rise in
sea level.

It is tempting to consider the coarser deposits reported by Mohr and
White from the central parts of the Java Sea as Pleistocene sediments
deposited subaerially during the glacials when chemical weathering was
less intense and coarser material, lithology permitting, thus could be
expected. They may, however, in part or completely form part of an older
planation surface also. Since the Pleistocene deposits studied by Biswas
are mainly clayey they do not give an answer to this question. The profiles,
given by Aleva et al are of interest in this relation: the Upper Tertiary to
Pleistocene Alluvial Complex underlying the Holocene/Recent clay blanket
is largely composed of sandy layers with some clayey and peaty intercalations. It occurs in the form of an extensive and complex system of valleys
and depressions incised in the Tertiary "Older Sedimentary Cover" (Fig.
11).

The sections through the sequence of Pleistocene deposits found in various localities in Java (von Koenigswald; 1940; Smit Sibinga, 1949, 1952)
seem to support the view that interglacial transgressions were accompanied
by fine-textured, clayey deposits and alternate with coarsertextured materials (sandstones, conglomerates, etc.) dating from glacial regression
periods.

Smit Sibinga has interpreted these characteristics of the Quaternary
stratigraphy of Java and also of Sumatra and Borneo as a result of the
glacio-eustatic changes in sea level. This view is not shared by the author
as will be explained in Section 4. 2.

4. 2   Coastal changes

The drop in sea level during the Pleistocene glacials caused the emergence
of vast tracts of low relief in the areas at present occupied by the Sunda
and Sahul shelfs. The drainage pattern then existing in those parts is a
matter of considerable interest and was studied first on the present Sunda
shelf by Molengraaff and Weber (1920). It was found that a huge North
Sunda river flowing northward in the present South China Sea collected its
tributaries from eastern Sumatra, western Borneo and from Malaya and
other parts of the east coast of mainland Asia. The drainage of the present
Java sea was directed eastward but the drainage lines originating in southeast Sumatra and northern Java on the one hand were largely separated
from those coming from southern Borneo by a gentle eastwest stretching
ridge situated in the central parts of the Java Sea. A much smaller river
drained northwestward between Malaya and northern Sumatra. The off
shore continuation of numerous rivers has been traced on the Singkep Islands and other nearby tin producing islands and is of economical importance because of workable placer tin deposits being associated with them
(Molengraaff and Weber, 1920; Aleva et al, 1973).

The drainage of the Sahul shelf has been unravelled more recently. The
drainage of this area has been dominantly westward, the Torres Strait

acting as a divide in the east. Wallace (1857), on the strength of the occurrence of sinuous channels ("sungi") between the islands of the Aru Archipelago situated near the western edge of the Sahul shelf, maintains that the main drainage line during the Pleistocene low sea levels has passed across the Aru Islands. Fairbridge (1953), however pointed out that the east-west stretching Merauke Ridge, stretching from the south coast of New Guinea to the Aru Islands prevented the New Guinea rivers from reaching the southern part of the Sahul shelf and from crossing the Aru Islands. Verstappen (1959) stressed that the channels ("sungi") between the Aru Islands are related to jointing and adheres to Fairbridge's views. Two separate rivers thus existed on the Sahul shelf during the Pleistocene, both flowing westward and situated in the Digul-Fly depression north of the Merauke Ridge and south of this ridge near the north coast of Australia respectively. The drainage pattern on the Sunda and Sahul shelfs during the Pleistocene low sea levels is indicated in Fig. 6.

An important question is whether or not the lowering in sea level resulted in incision of the river courses. Smit Sibinga (Baartmans et al., 1947; 1949; 1951) with Lehmann (1936) is of the opinion that the inter- and postglacial high sea levels provoked valley fill that was dissected when a lowering in sea level followed. The low terrace is believed by them to have formed during the highest postglacial sea level, and the high terrace found along the rivers in eastern Java is associated with the Risz-Würm interglacial. Two objections may be raised against this concept of valley development. Firstly, the earlier-mentioned "clay blankets" formed during interglacial high sea levels and humid tropical conditions are located near the

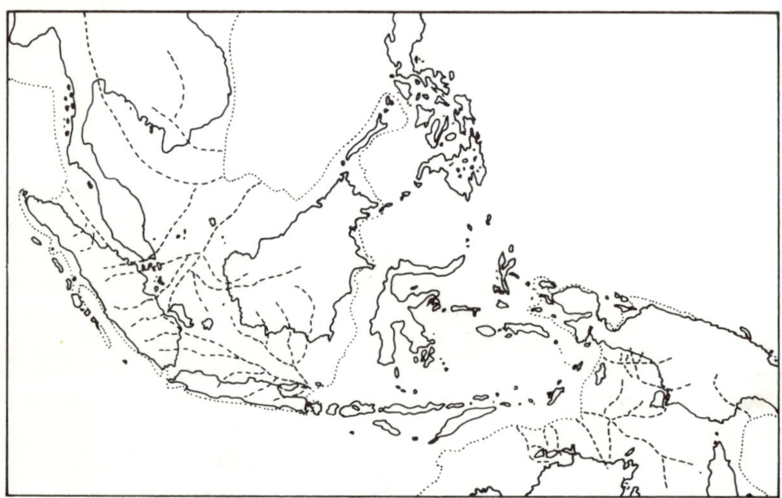

Figure 6. The great distance existing between the present coastlines and the edges (dotted lines) of the Sunda and Sabul shelfs is a measure for the great areas that became dry during the Pleistocene glacials. The major drainage lines then existing on the shelfs are indicated by dashed lines.

interglacial coast lines. They are, in part, deposited off-shore in the shallow coastal waters and partly as subaerial deposits in an alluvial plain or deltaic environment. These deposits, however, do not necessarily imply that deposition also dominated the more upstream parts of the river courses.

A second objection that one may raise is the fact that the glacial extensions of the lower river courses in the shelf areas had an extremely gentle gradient and thus did not invite incision of the riverbeds. This applies particularly to those areas drained by the huge river systems of the Sunda and Sahul shelfs. The present author therefore renounces the doctrine of glacial incision of river courses, and, for other reasons, elaborated upon in Section 4.3, is rather inclined to generally link the incision, in areas that remained emerged, to the interglacial conditions.

Nevertheless it goes to the credit of Smit Sibinga that he was the first who associated the alternation of marine and terrestrial Pleistocene sediments on Java, Sumatra (1947; 1949; 1951) and Borneo (1953) with factors other than tectonic movements ("oscillations") as had been advocated previously by Duyfjes (1938) and others. The coastal displacements in northeastern Java assumed by Smit Sibinga are given in Fig. 7 as an example. Much stress is put by Smit Sibinga on the alternation of interglacial clayey deposits and coarser textured glacial deposits. The present author

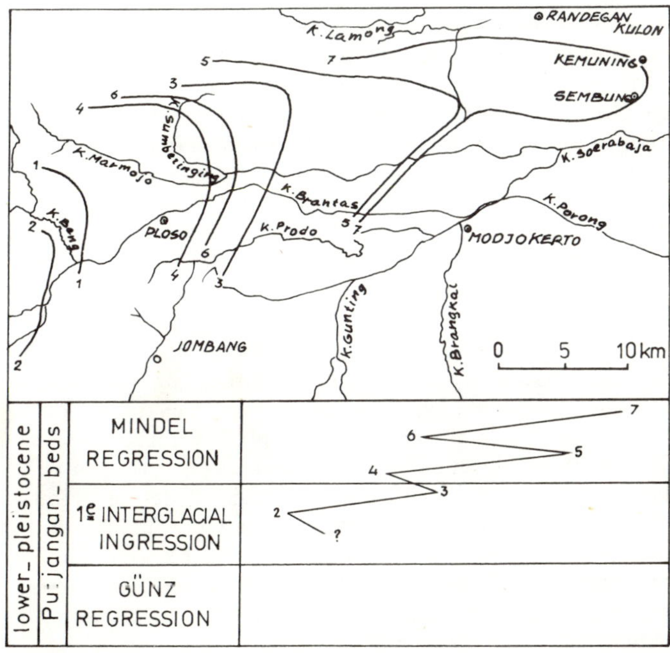

lower – pleistocene	Pudjangan – beds	MINDEL REGRESSION			
					7
			6		5
			4	3	
		1ᵉ INTERGLACIAL INGRESSION	2	?	
		GÜNZ REGRESSION			

Figure 7. Changes in the position of the coastline in part of East Java as a result of glacio-eustatic changes in sealevel during the Pleistocene, according to Smit Sibinga (1952).

agrees with this but considers climatic changes as a main cause of these alternations, as has been explained earlier in this paper.

The glacial chronology of Java and Sumatra given by Smit Sibinga certainly needs thorough revision also in the light of the few absolute datings that have been obtained since. The upper and lower limits of the Trinil fauna, (Kabuh and Djombang formations) have been K-Ar dated as 595.10^3 and 723.10^3 years respectively (Von Koenigswald, 1968; Jacob, 1972). The dating (700.10^3 years) of the Indoaustralian tektites "rain" thought to be contemporaneous with the age of the Pithecanthropus (Jacob, 1972), concord with this. Jacob also mentions the K-Ar dating of $1,900.10^3$ years ($\pm 400.10^3$ years) of the Djetis fauna (Putjangan formation) near Modjokerto a few meters below the site of the Modjokerto skull. The time-scale has been stretched considerably and it is thus questionable whether the high-terrace deposits (with Ngandong fauna) indeed date from the Riss-Würm interglacial as Lehmann and Smit Sibinga maintain, taking into consideration that the upper limit of the underlying Trinil fauna dates 595.10^3 years back!

Also more recent climatic-geomorphological investigations, such as the ones by Hopley (1973) on clay blankets and pediments in northeast Queensland, Australia, are bedevilled by the concept of river incision during glacial low sea levels.

Aleva et al. (1973) mention the occurrence of extensive systems of complicated valleys and depressions with a mostly sandy fill and local peat intercalations at depths of up to 60-100 meters below present sea level. This "alluvial complex" is of Upper Tertiary to Pleistocene age. No specification as to the glacial or interglacial age of the valley fill is given. It is incised in the Tertiary Older Sedimentary cover and overlain by a planation surface, which according to them, dates from the Risz-Würm interglacial and is formed by marine abrasion. The author is of the opinion that it was formed sub-aerially during low sea levels and that the role of abrasion in forming planation surfaces in Malesia has been insignificant because of the protected character of many coasts and the usually weakly developed waves which are only of importance where the Indian Ocean washes the shores and at some other localities. Also at present the geomorphological effect of wave action is strictly localized and of subordinate importance.

An interesting but much neglected aspect of Quaternary coastal development in SE Asia, and ranking third after changes in sea level and changes in texture of littoral sediments, is the change in coastal configuration induced by the variations in azimuth of the dominant wind direction, resulting from the shift in average position of the ITC. Verstappen (1953a, b, 1954b, 1968) demonstrated important fluctuations in dominant wind direction concurrent with shifts in position of the ITC in Jakarta since the beginning of this century that governed the development of shingle ramparts on the coral reefs situated in Jakarta Bay. The ramparts develop on the windward side of the coral reefs and subsequently are slowly pushed to the sand cays on the reef flat. Fig. 8 clearly illustrates the changes in the wind-pattern and Figs. 9 and 10 give the concurrent changes in position of the

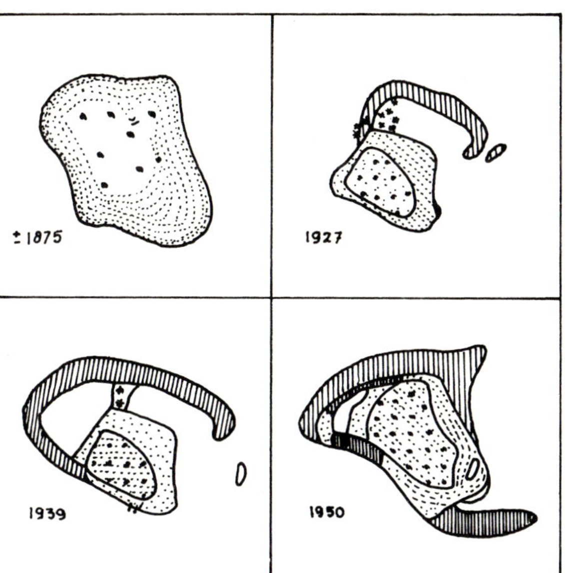

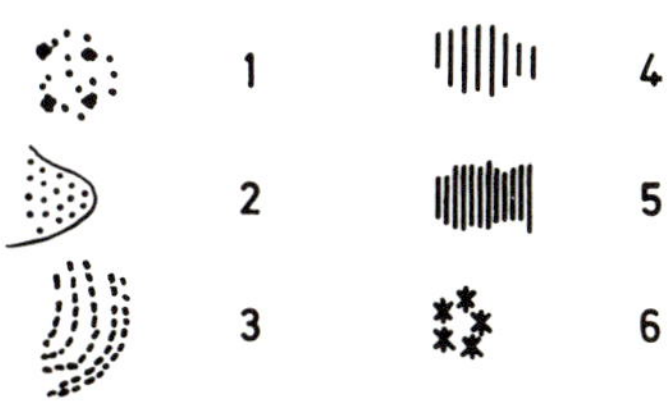

Figure 9. The island Air (Haarlem) as it appeared in 1875, 1927, 1939 and 1950. (Verstappen, 1954).

1. old nucleus;
2. recent accretion (on the 1875 map also shingle ramparts);
3. lines of growth;
4. low shingle ramparts (or sparse shingle);
5. high shingle ramparts;
6. isolated mangroves.
Scale 1 : 8,000

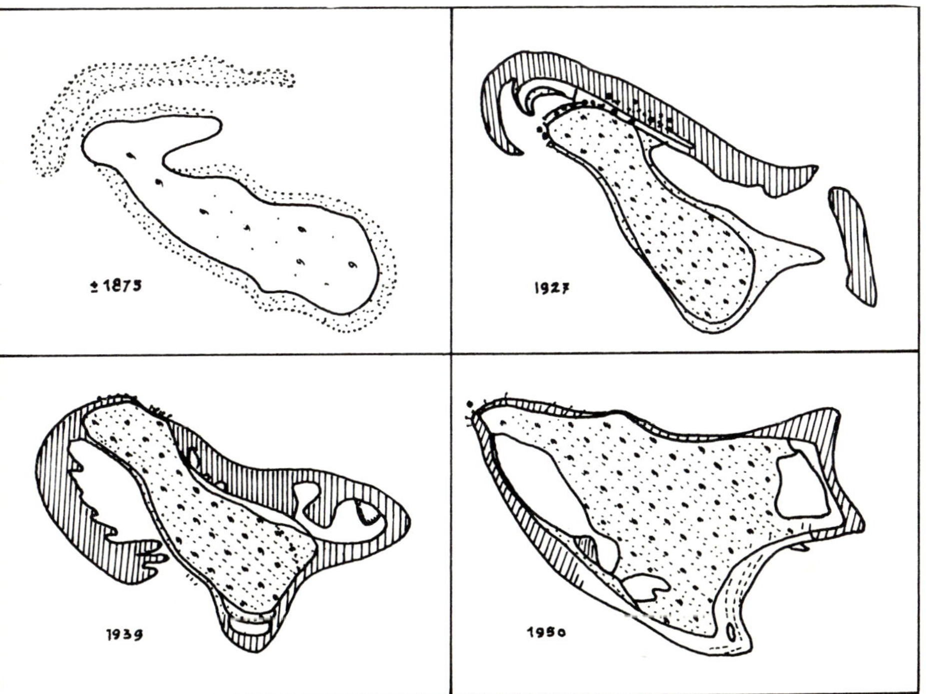

Figure 10. The island Njamuk Besar (Leiden) as it appeared in 1875, 1927, 1939 and 1950. Scale 1 : 8,000 (Verstappen, 1954).

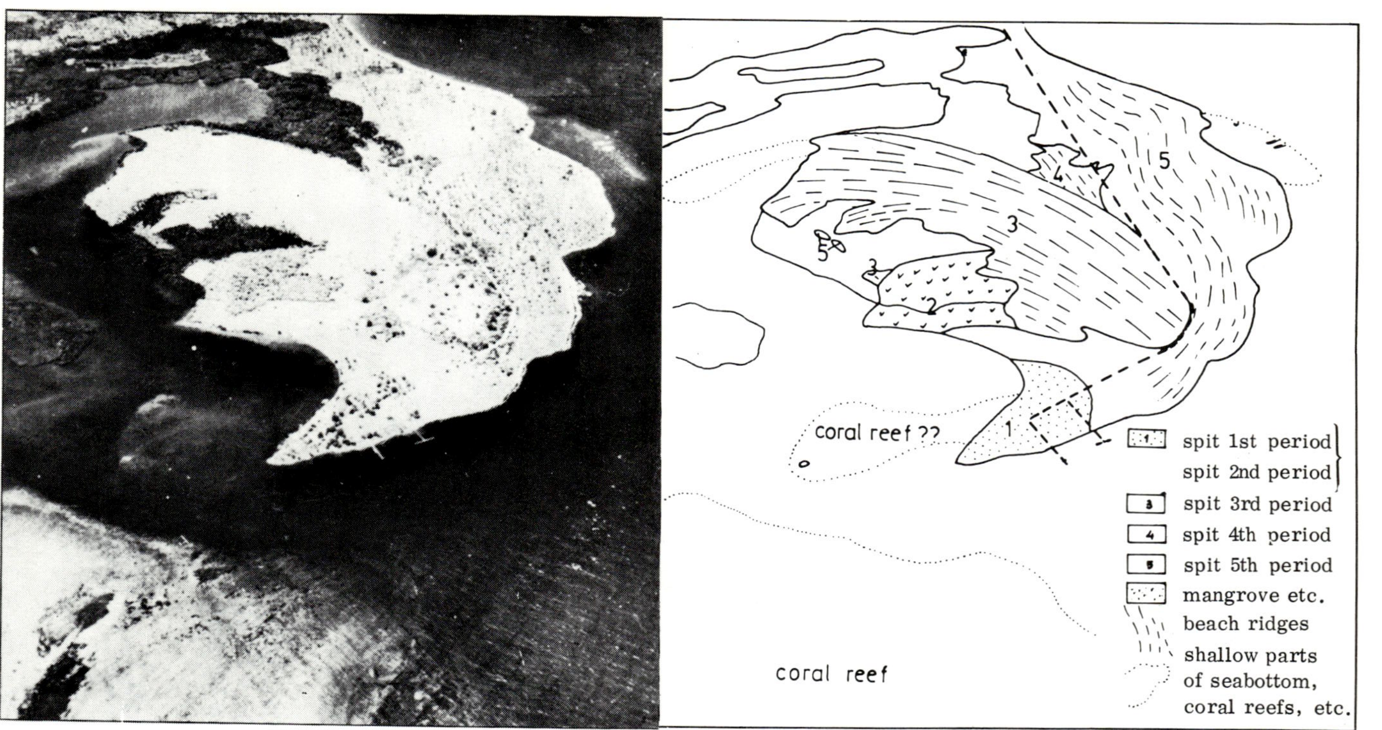

Figure 11. Oblique aerial view, with interpretation, of the Gilimanuk compound spit, located in the western extremity of the island of Bali, Indonesia. Five phases of development of the spit can be distinguished, each of which, according to the author, is associated with a period of strong winds from westerly directions (right side and upper right corner of the photo) and thus of a northern position of the ITC when the Asiatic Anticyclone was comparatively weakly developed. A C-14 dating of a prehistoric site situated between Phase 1 and 2 is expected (Soejono, 1973).

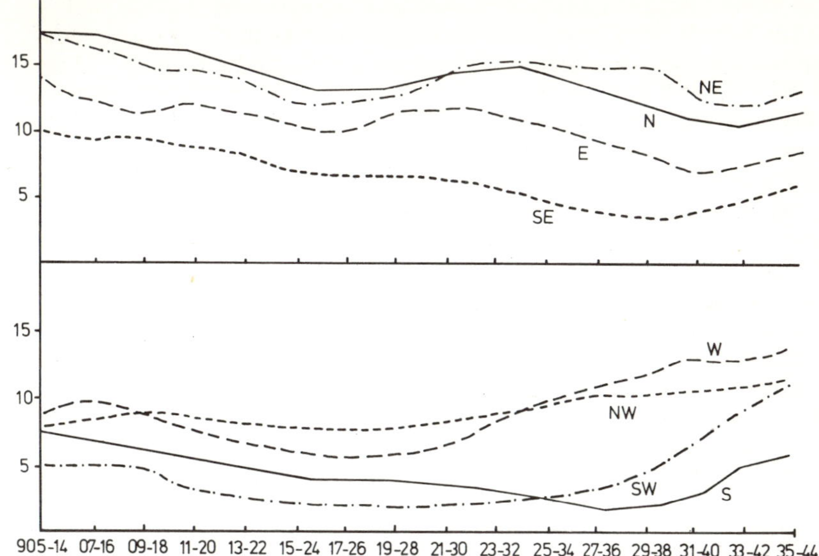

Figure 8. The 10-year sums (moving averages) of the product of the frequency (F) of the wind in hours per year and the mean wind velocity (V) in metres per second for the eight wind directions in the period 1905/14 - 1935/44 (Verstappen, 1954).

shingle ramparts. The same processes also affect other coastal zones, particularly lowland coasts. The dominant winds during the wet season, when the sediment yield of the rivers is maximum, determine to a large extent the direction of longshore currents which govern the place of deposition of clayey and sandy materials. The latter will form true (sandy) beach ridges particularly if the onshore winds are sufficiently forceful to produce an uprush which causes the sand to be deposited well above the waterline.

The position and periodicity of the beach ridges etc formed since the postglacial climatic optimum ($\pm$ 5000 BP) offer a clue to unravelling the climatic fluctuations that have occurred in that period. The complex spit of Gilimanuk (Bali) given as an example (Fig.11), evidences 4–5 periods of strong westerly winds (northerly position of ITC) in the last few thousand years.

It stands to reason that these processes also operated during the Pleistocene low sea level and that consequently the pattern of beach ridges then formed deviated from the present pattern. Since the load of the rivers in many cases must have been distinctly coarser-textured than at present, due to the less intense chemical weathering, the beach ridges were probably also better developed than during the interglacial high sea levels when the humid tropical weathering produced dominantly clays.

The wind effect also must have played an important role in the distribution of mangrove coasts. As mangroves flourish best along sheltered

23

parts of the coast, a change in dominant wind direction almost inevitably
affects the habitats of mangroves. It is of interest in this context to mention the widespread occurrence of mangrove pollen dating from the Pleistocene regressional phases in cores off the east coast of Malaya, which is
in strong contrast to their scarcity in Holocene sediments (Biswas, 1973).
At present mangrove forests only occupy 1.5% of this coastline.

A few words should be said about the effect of Pleistocene glacials on
coral reef development. A drop in sea water temperature of 4-5^{0}C, though
narrowing the coral belt in a latitudinal sense, still leaves the area under
consideration well above the 18^{0}C limit often set as a limit to coral up-
growth. The sea water near many coastal areas probably was less turbid
during the glacials because of the less intense weathering which fact is even
favourable for reef development. Evidently the coral growth was inter-
rupted in the shelf seas that ran dry. The reefs there could only grow
during the interglacials and the present reefs are all developed during
and after the postglacial rise in sea level. They therefore differ in type
size and mode of development from the reefs found in the deep seas.

4.3   Changes in morphogenetic development

This paragraph deals in particular with the mode of planation in those
cratogene and tectogene areas in Malesia where the Quaternary climatic
changes resulted in repeated alternation of tree savannah conditions and
humid tropical environment. Subaerial planation by the author is considered an important process during the tree savannah conditions that characterized the glacials. whereas linear river erosion characterizes the interglacial and Holocene periods.

This concept is based on an evaluation of the geomorphological processes that operated in the various parts of the Quaternary as discussed in
Section 4.1. The substantially drier, and somewhat cooler, climate that
prevailed in SE Asia during the Pleistocene glacials decreased the intensity of chemical weathering to some extent and so caused a coarser load
of many rivers which had a strongly seasonal regimen because of the pro-
longed dry season. The decreased, mainly tree-savannah, vegetation that
occurred favoured unchannelled surface wash, particularly on more gentle
slopes and plains. The glacial periods (and possibly also the Villafranchian) therefore were characterized by the building up of valley fill deposits
and alluvial fans. Large tracts, particularly in the at present immerged
parts of the shelf seas were covered by mostly sandy deposits.

The coarse surface deposits in parts of the Java Sea mentioned by Mohr
(1919) and the largely sandy Pleistocene "alluvial complex" described by
Aleva et al. (1973), may find their explanation in the conditions just mentioned. The process of double planation in the sense of Büdel must have
optimally developed with the chemical weathering still being strong enough
to plain off the bedrock and the open vegetation permitting planation at the
surface by non-concentrated wash. The "Younger Planation Surface" cap-
ping the Alluvial complex and the older formations (Fig.12) by Aleva et al.

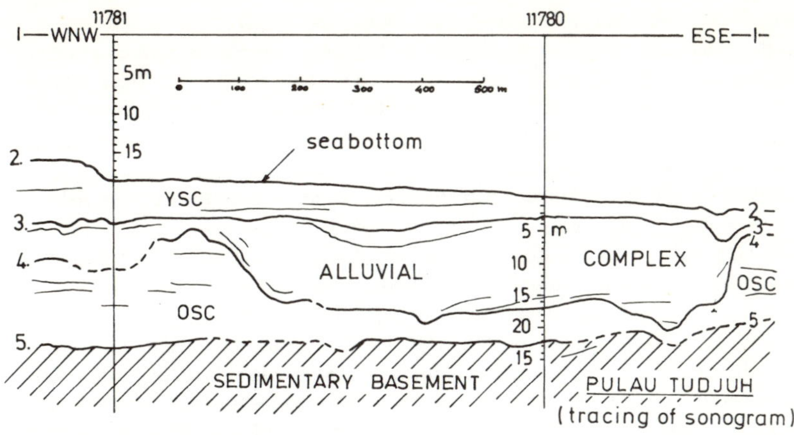

Figure 12. Tracing of sonogram showing the relationship between the alluvial complex and the sedimentary covers near Tudjuh Island (Aleva et al, 1973).

  Key:

YSC: young sedimentary cover

OSC: older sedimentary cover

1. sea surface
2. sea bottom
3. upper planation surface (with red clays)
4. base of alluvial complex
5. lower erosional surface

are interpreted as an abrasion platform developed during the Riss-Würm interglacial. According to the present author it may well be formed by subaerial planation due to non-concentrated surface wash under the tree savannah conditions of the Würm glacial. The red clay formation dating from the Würm glacial observed by them on top, and also noted by Biswas (1973), fits logically in the concept of subaerial origin of the Younger Planation Surface.

Another feature that may be explained in the same way is the so-called "peneplain" of Palembang in southern Sumatra, described by Boissevain (Baartmans et al, 1947), Smit Sibinga (1951) and Tobler (1912), the origin of which is a long discussed subject. A comparable problem also exists in the low areas of northern Sumatra (Zwierzycki, 1920). It is unthinkable that planation surfaces of this size can be formed under the conditions at present prevailing with rivers confined to their beds by the dense tropical vegetation and linearly incising. Tobler (1912), with justification, already distinguished between the high terrace situated in valleys and following the course of the major rivers in southern Sumatra and high terrace deposits covering the ridges between the present rivers. The process was not only restricted to planar sedimentation, but also comprised of planing of older formations.

Also observations have been made on Borneo that seem to concord with the view of planations during the Pleistocene glacials expressed in this paper. The "peneplain" of Kutei and Bulongan studied by L. Rutten (1927) and considered by him as dating from the Plio-Pleistocene according to de Sitter (1948) is much younger and was formed in the Late Pleistocene. Its dissection is largely attributed to tectonics. Deformation is assumed since anticlinal axes and synclinal depressions can be traced. If, however, surface wash is accepted as a main factor of planation resulting in planing off, in the hills and simultaneously in accumulation at the hill feet, such topographical irregularities are readily explained, as is the sandy/gravelly nature of the deposits found on top of this planation surface.

Interesting observations have been made in Malaya where Walker (1956; see also Ingham and Bradford, 1960) made a detailed study of the so-called Old(er) Alluvium in the Kinta Valley, Perak. He mentions the occurrence of Boulder Beds in discontinuous strips along valley sides and as infill of small tributary valley heads and of more extensive deposits of Old Alluvium which are mainly composed of horizontally bedded clays with sand and gravel laminae.

The Old Alluvium is aggraded up to 70 meters above present sea level and concordant with a horizontal terrace formed in the Boulder Beds and in solid rock. The consistent grainsize and the horizontality of the terrace have led Walker to the conclusion that the deposits are of marine origin and were formed when the sea level was about 70 meters higher than to date. Two lower terraces are also reported, at 15-30 and 2-5 meters m. s.l. respectively. They are, likewise, believed to be of eustatic origin. The wide valleys, cut in bed rock that have been filled up with alluvium in the past and now contain only relatively small meandering streams, according to Walker, are a "drowned topography". He associates the incision with low sea levels and the infill to periods of higher sea level. Only recently (Sivam 1968; 1969; Newell, 1971) a fluvial origin of the deposits has been established. Confusion still exists, however, and incision is associated with low Pleistocene sea levels (Gobbett and Hutchison, 1973).

Burton (1964), who studied the Older Alluvium of Johore and Singapore, describes it as clay, sand and gravel, ranging in altitude, from 45 meters below sealevel to 75 meters above sea level. The deposits can be distinguished from the Younger Alluvium because they are laterized, have a deep weathering profile, are in a semi-consolidated state and are dissected to low hills. Like Walker he believes in a dominantly marine origin associated with a sea level 75 meters above the present one. It is worth mentioning in this context that Wall (1967) attributes deposits in northern Borneo, which are probably equivalent to the Old Alluvium, to Quaternary high sea levels. A slightly different opinion is expressed by Ashton (1972) who advocates a Mid Pleistocene age of the Old Alluvium on the strength of tektites, estimated at 700,000 BP, occurring at their base. He believes that the deposits were formed during relatively short periods at the end of the Günz, Mindel and Riss glaciations.

It appears to the author that the Old Alluvium is much more easily accounted for without the necessity of postulating higher sea levels if the

concept of wash deposits and colluvials formed independent of sea level
under (tree) savannah conditions during the Pleistocene glacials is accepted. The just mentioned rock-cut terrace, reported by Walker from the
Kinta Valley, is then obviously a tropical foot plain formed under savannah
conditions as outlined earlier in this paper. These foot plains are rather
wide spread in Malaya and have been excellently described by Nossin
(1964), who at the time did not give thought to the different climates that
have prevailed in the past. Pediments and associated features have
recently also been reported from northeast Queensland, Australia and are
considered evidence for drier conditions that are supposed to have prevailed there during the Pleistocene glacials (Hopley, 1973). Interglacial
clay blankets are also mentioned. These observations tie in well with the
thoughts developed in this paper, except for the concept of low sea level
river incision adhered to by Hopley.

The author's views find support in the investigations carried out by
Haile and Ayob of the Quaternary sediments in the Sungei Besi open
cast tin mine near Kuala Lumpur. These deposits unconformably overlie
granites and limestones and have their base below present sea level. Three
C-14 datings near the base give an age of 36, 420; 41, 200 (minimum) and
41, 500 (minimum) BP respectively. Although the reliability of C-14 dating
of deposits of this age is questionable, the evidence suggests that the
deposits were formed during the low sea levels of the Würm (and earlier?)
glacial(s). Ayob (1970), with justification, says that the lack of marine
fossils and the occurrence of plant remains and gravelly layers suggests a
fluvial origin and that the dominance of angular quartz grains, tourmaline
and muscovite indicates nearby granitic rocks as a source. Wash from
the hills thus seems to have played an important role in the sedimentation
process.

The age determination of the Older Alluvium is also attempted by Sivam
(1968, 1969) in the Kinta Valley where C-14 datings of 39, 000 and 39, 000
BP as compared to 3, 070 ± 100 BP for the Younger Alluvium were obtained.
It seems likely to the author that these and comparable deposits have been
formed during the Würm and earlier Quaternary glacials and since then
may have been in part removed. It goes to the credit of Haile (1971) to
have completely done away with Quaternary high sea levels (other than an
approximately 6 m Holocene sea level) that so long have bedevilled Quaternary research in West Malaysia.

The humid tropical conditions characterising the interglacial periods
and the Holocene resulted in the deposition of a clay blanket near the (high)
coastlines both above and below the waterline. Further upstream, however,
the linear riverwork resulted in incision of the rivers. Chemical weathering became more intense and thus the planation at depth by alteration of the
bedrock was maximum, at least in areas of low relief. The surface wash,
however, was almost completely stopped and thus the process of "double"
planation became ineffective. After incision, sliding in the thick weathering mantle and other processes produce a different mode of landform
development. In the long run this possibly also leads to a sort of planation

which cannot be compared with savannah (double) planation, however, since non-channelled surface wash as a factor of surfacial planation is replaced by the removal of clays to the drainage lines.

A final remark should be made regarding the different development, on the one hand in (cratogene) equatorial areas of low relief where conditions for planation are optimum and have been studied by many authors and, on the other hand, (usually tectogene) equatorial areas of higher relief which have received much less attention. In the latter, under humid tropical conditions not only clay is carried off in suspension, but gravel and boulders are normal occurrences in the river beds. Planation is of local importance only, since the rivers are mostly incising their courses. When the tropical rainforest is replaced by a less dense vegetation following a decrease in rainfall and humidity, the thick weathering mantle previously formed will be readily removed and the debris will be deposited at the foot of the hills, forming fans and generating footplains that to some extent resemble pediments and glacis known from more arid regions. They differ from these, however, because of their linkage to relief or tectonics rather than to lithology only, and, (except in alpine mountains) by the relative scarcity of coarse angular debris, characteristic of many foot plains in drier areas. The greater humidity and associated chemical weathering renders a rapid foot-planation of this type possible and explains the dominance of fine and medium textured material involved in the process.

Towards the lowlands these foot plains, topography permitting, may give way to extensive "double" planation surfaces. These are best developed in those parts of lowland Malesia where the vegetation was most drastically reduced e.g. in areas that even nowadays have a marked dry season and comparatively low rainfall. Towards the hills these foot slopes end rather abruptly in the much steeper hill slopes which are partly exhumed. Thus exfoliation domes and tors with surrounding footslopes are easily developed — lithology permitting — and have been reported e.g. from Bangka and Belitung by Faber (1955a, b) and from Malaya by Nossin (1964). The formation has been prepared by intense chemical weathering, particularly along joints under humid-tropical conditions; surface wash under drier conditions furthered their exhumation and caused the footplains surrounding them.

5 CONCLUSIONS

It can be concluded that drier conditions with lower precipitation values and a longer dry season, have occurred in Malesia during the Pleistocene glacials. They had an important effect on vegetation and also on landform development. Planation and deposition of — lithology permitting — mostly sandy materials in the areas of low relief and the formation of valley fill and fans in higher parts is characteristic for the glacial periods. The low sea levels of these periods are believed generally not to have caused river incision since the downstream, presently immerged, river sections had a very low gradient.

Incision is mainly tied to the interglacial and Holocene humid tropical
conditions when vegetation interfered with non-concentrated surface wash.
Deposition of a clay blanket is effectuated in these periods near the (high)
sea level both above and below the waterline. The Quaternary chronology
of Malesia will have to be revised in view of this concept.

Attention is drawn to the often neglected and different developments in
mountainous (tectogene) equatorial areas of high relief as compared to
(cratogene) zones of lower relief. Extensive "double" planation is thought
to have been effective only during the glacial periods in the drier areas of
low relief. It never played a role in the areas of higher relief since incision here was dominant all the time. The thick weathering mantle under
tropical rain forest conditions here was affected by mass movements and
removal of clays. When drier conditions invaded these areas of higher
relief, the vegetative cover was reduced and the mantle was effected by
surface wash which process resulted in the (near) exhumation of the
rocks on many slopes and in the formation of tropical foot slopes and fans.
These conditions also favoured the exhumation of exfoliation domes and
tors reported from parts in the area.

6 BIBLIOGRAPHY

Aleva, G.J.J. et al, 1973. A contribution to the geology of part of the
Indonesian tinbelt: the sea areas between Singkep and Bangka islands
and around the Karimata islands. Proc. reg. Conf. Geol. SE Asia.
Bull. geol. Soc. Malaysia, 6: 257–271.

Andel, Tj.H. van & J.J. Weevers, 1965. Submarine morphology of the
Sahul shelf, northwestern Australia. Bull. geol. Soc. Amer., 76: 695–
700.

Ashton, P.S., 1972. The Quaternary geomorphological history of western
Malesia and lowland forest phytogeography. Trans. Aberdeen–Hull
Symp. malesian ecology. Univ. Hull, Dept. Geogr. Misc. Ser., 13: 35–
49.

Ayob, M., 1970. Quaternary sediments at Sungei Besi, West Malaysia.
Bull. geol. Soc. Malaysia, 3: 53–61.

Baartmans, J.A. et al, 1947. De morfologie van de Java- en Soenda zee.
Tijdschr. K. Ned. aardrijksk. Gen., 64: 442–465; 555–576.

Baren, F.A. van & H. Kiel, 1950. Contribution to the sedimentary petrology of the Sunda shelf. J. Sedim. Petrology, 20: 185–213.

Bik, M.J.J., 1967. Structural geomorphology and morphoclimatic zonation
in the Central Highlands, Australian New Guinea. Pages 26–47 in: Landform
Studies from Australia and New Guinea. J.N. Jennings & J.A. Mabbut
(eds), Canberra, ANU Press.

Biswas, B., 1973. Quaternary changes in sea-level in the South China Sea.
Proc. reg. Conf. Geol. SE Asia. Bull. geol. Soc. Malaysia, 6: 229–255.

Braak, J., 1923. Het klimaat van Nederlandsch-Indië.Verh. kon. magn.
met. Obs. Batavia, 1.

Bremer, H., 1967. Zur Morphologie von Zentral-australien. Heidelb.
  geogr. Arb., 17.
Büdel, J.K., 1957. The ice age in the tropics. Universitas, 1: 183-192.
Burkill, I.H. & R.E. Holttum, 1923. Gard. Bull. Str. Settlem. 3: 19-23.
Burton, C.K., 1964. The older alluvium of Johore and Singapore. J. trop.
  Geogr., 18: 30-42.
Charlesworth, J.K., 1957. The Quaternary era with special reference to
  its glaciation. London, E. Arnold.
Coaldrake, J.E., 1968. Quaternary climates in northeastern Australia
  and some of their effects. Proc. R. Soc. Queensland, 79: 5-18.
Conolly, J.R., 1967. Postglacial-glacial change in climate in the Indian
  Ocean. Nature, 214: 873-875.
Crocker, R.L., 1959. Past climatic fluctuations and their influence upon
  Australian vegetation. Pages 283-290 in A. Keast et al (eds), Bio-
  geography and Ecology in Australia.
Derbyshire, E., 1969. Approche synoptique de la circulation du dernier
  maximum glaciaire dans le Sud-Est de l'Australie. Revue Géogr. phys.
  Géol. dyn., 11: 341-362.
Dozy, J.J., 1938. Eine Gletscherwelt in Niederländische Neu Guinea, Z.
  Gletscherkunde, 26: 45-51.
Dubois, E., 1890. Voorlopig bericht omtrent het onderzoek naar de pleis-
  tocene en tertiare vertebraten-fauna van Sumatra en Java gedurende
  het jaar 1890. Nat. Tijdschr. Ned. Indië, 51: 93-100.
Dubois, E., 1937. Early man in Java and Pithecanthropus erectus. Pages
  315-322 in Early Man, Philadelphia.
Duyfjes, J., 1938. Geologische kaart van Java. Toelichtingen bij de bladen
  110, 115, 116 (Modjokerto, Soerabaja, Sidoardjo). Dienst. Mijnbouw,
  Ned. Indië.
Emiliani, C., 1955. Pleistocene temperatures. J. Geol., 63: 538-578.
Emiliani, C., 1961. Cenozoic climatic changes as indicated by the strat-
  igraphy and chronology of deep-sea cores of globigerina ooze facies.
  NY Acad. Sci. Ann. 1, 95: 521-536.
Emiliani, C., 1966a. Isotopic palaeotemperatures. Science, NY, 154: 851-
  857.
Emiliani, C., 1966b. Palaeotemperature analysis of Caribbean cores
  P.6304-8 and P.6304-9 and a generalized temperature curve for the
  past 425,000 years. J. Geol., 74: 109-126.
Emiliani, C. & G. Wollin, 1956. Micropalaeontological and isotopic deter-
  minations of Pleistocene climate. Micropalaeontology, 2: 257-270.
Ericson, D.B. et al, 1961. Atlantic deep sea sediment cores. Bull. geol.
  Soc. Amer., 72: 193-286.
Faber, D.A., 1955a. Verslag van een bodemkundige verkenning van het
  eiland Bangka. Report Balai Penjelidikan Tanah, Bogor.
Faber, D.A., 1955b. Verslag van een bodemkundige verkenning van het
  eiland Billiton. Report Balai Penjelidikan Tanah, Bogor.
Fairbridge, R.W., 1953. The Sahul Shelf, Northern Australia: Its struc-
  ture and geological relationships. J. R. Soc. W. Aust., 37: 1-33.

Fairbridge, R.W., 1964. African ice-age aridity. Proc. NATO Conf. on Palaeoclimates, 1963. Newcastle upon Tyne: 356–363.

Fairbridge, R.W., 1970. World palaeoclimatology of the Quaternary. Revue de Gèogr. phys. Géol. dyn., 12: 79–104.

Flenley, J.R., 1969. The vegetation of the Wabag region, New Guinea Highlands: a numerical study. J. Ecol., 57: 465–473.

Flint, R.F. & F. Brandtner, 1961. Climatic changes since the last interglacial. Amer. J. Sci., 259: 321–328.

Flohn, H., 1952. Allgemeine atmosphärische Zirkulation in der letzten Eiszeit. Geol. Rundschau, 40.

Galloway, R.W., 1965a. A note on world precipitation during the last glaciation. Eiszeitalter und Gegenwart, 16: 76–77.

Galloway, R.W., 1965b. Late Quaternary climates in Australia. J. Geol., 73: 603–618.

Galloway, R.W., 1967. Dating of shore features of Lake George, NS Wales, Aust. J. Sci., 29: 477.

Galloway, R.W., 1971. Evidence for Late Quaternary Climates, in Mulwany & Colson, 1971.

Galloway, R.W., 1973. Late Quaternary glaciation and periglacial phenomena in Australia and New Guinea. Palaeoecology Africa, 8: 127–138.

Gentilli, J., 1961. Quaternary climates of the Australian Region. Ann. NY Acad. Sci., 95: 465–501.

Gill, E.E., 1955. The Australian "Arid Period". Aust. J. Sci., 17: 204–206.

Gobbett, D.J. & C.S. Hutchison (eds.), 1973. Geology of the Malay Peninsula. New York, London, Wiley-Interscience. 438 pp.

Haile, N.S., 1969. Quaternary deposits and geomorphology of the Sunda shelf of Malaysian shores. Proc. INQUA Congr., Paris.

Haile, N.S., 1970. Radiocarbon dates of Holocene emergence and submergence in the Tambelan and Benguran Islands, Sunda Shelf, Indonesia. Bull. geol. Soc. Malaysia, 3: 135–137.

Haile, N.S., 1971. Quaternary shorelines in West Malaysia and adjacent parts of the Sunda shelf. Quaternaria, 15: 333–343.

Hooijer, D.A., 1949. Mammalian evolution in the Quaternary of southern and eastern Asia. Evolution, 3: 125–128.

Hooijer, D.A., 1968. The Middle Pleistocene fauna of Java. Pages 86–90 in Evolution and Hominisation. G. Kurth (ed.), Stuttgart, G.G. Fischer.

Hoopen, K.J. ten & F.H. Schmidt, 1950. Recent climatic variations in Indonesia. Nature, 47.

Hopley, D., 1969/70. A coastal weathering sequence at Mt Douglas, North Queensland. Revue Geom. Dyn., 19: 9–14.

Hopley, D., 1973. Geomorphic evidence for climatic change in the Late Quat of NE Queensland, Austr. J. trop. Geogr., 36: 20–30.

Ingham, F.J. & F.F. Bradford, 1960. The geology and mineral resources of the Kinta Valley, Perak. Fed. Malaya geol. Survey. Distr. Memoir, 9: 347.

Jacob, T., 1972. The absolute date of the Djetis beds at Modjokerto. Antiquity, 46: 148.

Keller, H.K. & A.F. Richards, 1967. Sediments of the Malacca Strait, Southeast Asia. J. Sedim. Petrology, 37: 102–127.

Kershaw, A.P., 1970. A pollen diagram from Lake Enramo, Northeast Queensland, New Phytologist, 69: 785–805.

Kershaw, A.P., 1971. A pollen diagram from Quincan Crater, Norheast Queensland, New Phytologist, 70: 669–681.

Koeningswald, G.H. von, 1940. Neue Pithecanthropus-Funde 1936–1938. Wetensch. Meded. Dienst. Mijnbouw, 28: 232.

Koeningswald, G.H. von, 1950. Vertebrate stratigraphy. Pages 91–93 in R.W. van Bemmelen, Indonesia, Vol I.

Koeningswald, G.H. von, 1968. Das absolute Alter des Pithecanthropus erectus Dubois. Pages 195–203 in G. Kurth (ed.), Evolution and Hominisation. Stuttgart, G. Fischer.

Koopmans, B.N. & P.H. Stauffer, 1968. Glacial phenomena on Mt Kinabaluh, Sabah. Borneo Region, Malaysian geol. Survey Bull. 8: 25–35.

Kuenen, Ph., 1939. Sediments of the east indian archipelago. Rev. Marine Sediment, 3: 348–355.

Kraus, E.B., 1973. Comparison between ice age and present general circulation. Nature, 245: 129–133.

Lehmann, H., 1936. Morphologische Studien auf Java. Geogr. Abh. 3e Reihe, 9.

Löffler, E., 1970. Evidence of Pleistocene glaciation in East Papua. Aust. geogr. Stud., 9: 16–26.

Löffler, E., 1972. Pleistocene glaciation in Papua and New Guinea. Z. Geom., Suppl. Bd. 13: 46–72.

Mabbut, J.A., 1971. Denudation chronology in Central Australia. Structure, Climate and Landform Inheritance in the Alice Spring Area. Pages 144–181 in Jennings & Mabbut (eds.), Landform studies from Australia and New Guinea. Canberra, ANU Press.

Medway, Lord, 1964. Niah cave bone VII: size changes in the teeth of two rats, Rattus sabanus Thomas and R. muelleri Jentink. Sarawak Mus. J., 11: 616–623.

Medway, Lord, 1971. The Quaternary mammals of Malesia, a review. Trans. 2nd Aberdeen-Hull, Symp. malesian Ecology. Univ. Hull, Dept. Geogr. Misc. Series, 13: 63–98.

Mohr, E.Ch.J., 1900. Over het slibbezwaar van eenige rivieren in het Serajoedal en daarmede in verband staande onderzoekingen. Meded. Dept. Landbouw, 5. 95 pp.

Mohr, E.Ch.J., 1919. Sedimenten van de Java Zee. Verh. Ned. Indië Nat. Wet. Congr. Batavia: 219–223.

Mohr. E.Ch.J., 1922. De grond van Java en Sumatra. Amsterdam, 220 pp.

Molengraaff, G.A.F. & M. Weber, 1920. On the relation between the Pleistocene glacial period and the origin of the Sunda Sea (Java- and South China Sea), and its influence on the distribution of coral reefs and on the land- and fresh water fauna. Proc. R. Acad. Amsterdam, 23: 395–439.

Molengraaff, G.A.F., 1930a. The coral reefs in the East Indian Archipelago, their distribution and mode of development. Proc. 4th Pacif.Sci. Congr. Java, 2.

Molengraaff, G.A.F., 1930b. The recent sediments in the seas of the East Indian Archipelago. A short discussion on the conditions of those seas in former geological periods, Ibid.

Muller, J. 1972. Palynological evidence for change in geomorphology, climate and vegetation in the Mio-Pliocene of Malesia. Trans. 2nd Aberdeen-Hull Symp. malesian Ecology. Univ. Hull Dept. Geogr. Misc. Series, 13: 6–34.

Mulvaney, D.J. & J. Golson (eds.), 1971. Aboriginal Man and Environment in Australia. Canberra, ANU Press.

Musper, K.A.F.R., 1937. Geologische kaart van Sumatra. Toelichting bij blad 16 (Lahat). Dienst Mijnbouw Ned. Indië.

Neeb, G.A., 1943. Bottom samples. The composition and distribution of the samples. The Snellius Expedition, 5/3: 47–268.

Newell, R.A., 1971. Characteristics of the stanniferous alluvium in the southern Kinta Valley, West Malaysia. Bull. geol. Soc. Malaysia, 4: 15–37.

Nix, H.A. & J.D. Kalma, 1972. Climate as a dominant control in the biogeography of northern Australia and New Guinea. In Walker, 1972.

Nossin, J.J., 1964. Geomorphology of the surroundings of Kuantan (eastern Malaya). Geol. Mijnbouw, 43: 157–182.

Oba, T., 1967. Planktonic foraminifora from the deep-sea cores in the Indian Ocean. Tohoku Univ., Sendai, 2nd series, Geol. 38: 193–219.

Parker, F.L., 1962. Planktonic foraminiferal species in the Pacific sediments. Micropaleont., 8: 219–254.

Parker, F.L., 1967. Late Tertiary biostratigraphy (planktonic foraminifera) of tropical Indo-Pacific deep-sea cores. Amer. Palaeont. Bull. 53 (235). 208 pp.

Peterson, R.M., 1970. Würm II climate at Niah cave. Sarawak Mus. J., 17: 67–79.

Peterson, J.A. & G.S. Hope, 1972. A lower limit and maximum age for the last major advance of the Carstensz Glacier, West Irian. Nature, 240: 36–37.

Pfannenstiehl, M., 1953/54. Das Quartär der Levante, Wiesbaden.

Reiner, E., 1960. The glaciation of Mt Wilhelm, Austr. New Guinea. Geogr. Rev., 50: 491–503.

Rutten, L.M.R., 1917. Over denudatiesnelheid op Java. Kon. Akad. Wet. Amsterdam. Afd. Wisk. Nat., 26: 920–930.

Rutten, L.M.R., 1927. Voordrachten over de geologie van Nederlandsch Oost-Indië. Groningen, Wolters, 839 pp.

Rutten, M.R., 1952. Geosynclinal subsidence versus glacially controlled movements in Java and Sumatra. Geol. Mijnbouw, 14: 211–220.

Sartono, S., 1973. On Pleistocene migration routes of vertebrate fauna in southeast Asia. Proc. Reg. Conf. Geol. SE Asia. Bull. geol. Soc. Malaysia, 6: 273–286.

Schmidt, F.H. & K.J. Schmidt-ten Hoopen, 1951. On climatic variations in Indonesia. Djaw. Meteor. Geofis., Verh. 41, 43 pp.

Schmidt, F.H. & J.H.A. Ferguson, 1952. Rainfall types based on wet and
dry period ratios for Indonesia with western New Guinea. Djaw. Meteor.
Geofis., Verh. 42. 77 pp.

Scrivenor, J.B., 1949. Geological and geographical evidences for change
in sea level during ancient Malayan history and pre-history. J. Malay
Branch R. Asiatic Soc., 22: 107-115.

Sitter, L.U. de, 1948. Het Quartair in het kustgebied van Koetei ten
noorden van de Mahakam Rivier. Geol. Mijnbouw, 10: 177-183.

Sivam, S.P., 1968. Radiocarbon dating in the Kinta Valley. Geol. Soc.
Malaysia Newsletter, 15, 1.

Sivam, S.P., 1969. Quaternary alluvial deposits in the north Kinta Valley,
Perak, M.Sc. thesis Kuala Lumpur.

Smit Sibinga, G.L., 1949. Pleistocene eustasy and glacial chronology in
Java and Sumatra. Verh. K. Ned. geol. mijnbouw Gen., 15/1: 1-30.

Smit Sibinga, G.L., 1951. On the origin and age of the peneplain of Palembang. Geol. Mijnbouw, 13: 1-11.

Smit Sibinga, G.L., 1952. Interference of glacial eustasy with crustal
movements and rhythmic sedimentation in Java and Sumatra. Geol.
Mijnbouw, 14: 220-225.

Smit Sibinga, G.L., 1953. Pleistocene eustasy and glacial chronology in
Borneo. Geol. Mijnbouw, 15: 365-383.

Soejono, R.P., 1973. The significance of the excavations at Gilimanuk
(Bali). In press.

Stauffer, P.H., 1968. Glaciation in Mount Kinabalu. Geol. Soc. Malaysia
Bull. 1. 63 pp.

Steenis, C.G.G.J. van, 1935. On the origin of the Malaysian mountain
flora, Part 2, sér III, Bull. Jard. bot. Buitenzorg, 13: 398.

Steenis, C.G.G.J. van, 1939. The native country of sandalwood and teak.
A plant-geographical study. Hand. 8e Ned. Indië Nat. Wet. Congr.,
Soerabaja 1938: 408-409.

Steenis, C.G.G.J. van, 1947. Tertiary fossil wood from the islands
Soemba and Soembawa, and its genetic plant-geographical significance.
Chronica Naturae, 103: 1-2.

Steenis, C.G.G.J. van, 1961. Introduction. In M.S. van Meeuwen et al.
Preliminary revisions of some genera of Malaysian Papilionaceae, I.
Reinwardtia, 5: 420-429.

Steenis, C.G.G.J. van, 1965. Concise Plant-Geography of Java. Flora
of Java, Vol 2: 1-72.

Terra, H. de, 1943. Pleistocene geology and early man in Java. Trans.
Am. phil. Soc. N.S., 32: 437-464.

Tjia, H.D., 1966a. Arafura Sea. Encycl. of Oceanography: 44-47.

Tjia, H.D., 1966b. Java Sea. Encycl. of Oceanography: 424-429.

Tjia, H.D., 1970. Quaternary shorelines of the Sunda land, southeast
Asia. Geol. Mijnbouw, 49: 135-144.

Tobler, A., 1912. Geologie van het Goemai gebergte (Res Palembang,
Zuid Sumatra). Jrb. Mijnwezen Ned. Indië: 6-49.

Verstappen, H.Th., 1952. Luchtfotostudies over het Centrale Bergland van Nederlands Nieuw-Guinea. Tijdschr. K. Ned. aardr. Gen., 69: 336–363; 425–431.

Verstappen, H.Th., 1953a. Djakarta Bay, a geomorphological study on shore line development. Thesis, Utrecht Univ., 101 pp.

Verstappen, H.Th., 1953b. Oude en nieuwe onderzoekingen over de koraal eilanden in de baai van Djakarta. Tijdschr. K. Ned. aardr. Gen., 70: 472–478.

Verstappen, H.Th., 1954a. Het kustgebied van noordelijk West-Java op de luchtfoto. Tijdschr. K. Ned. aardr. Gen., 71: 146–152.

Verstappen, H.Th., 1954b. The influence of climatic changes on the formation of coral islands. Amer. J. Sci., 252: 428–435.

Verstappen, H.Th., 1955. Geomorphologische Notizen aus Indonesien. Erdkunde, 9: 134–144.

Verstappen, H.Th., 1959. Geomorphology and crustal movements of the Aru Islands in relation to the Pleistocene drainage of the Sahul shelf. Amer. J. Sci., 257: 491–502.

Verstappen, H.Th., 1960. Preliminary geomorphological results of the Star Mountains expedition 1959. Tijdschr. K. Ned. aardr. Gen., 77: 305–311.

Verstappen, H.Th., 1968. Coral reefs, wind and current growth control. Encycl. of Geomorphology: 197–202.

Walker, D., 1956. Studies on the Quaternary of the Malay Archipelago I. Alluvial deposits of Perak and the relative levels of land and sea. Fed. Mus. J., 1–2: 19–34.

Walker, D., 1970. The changing vegetation of the montane tropics, Search, 1: 217–221.

Walker, D., 1972. Bridge and barrier: the natural and cultural history of Torres Strait. Res. School of Pacif. Stud., Dept of Biogeogr. and Geom., Publ. BG/3. Canberra, ANU Press.

Wall, J.R.D., 1967. The Quaternary geomorphological history of North Serawak with special reference to the Subis Karst, Niah. Sar. Mus. J. (N.S.), 15, 30/31: 97–125.

Wallace, A.R., 1957. On the natural history of the Aru Islands. Ann. and Mg. Nat. Hist., ser 2, 20: 473–485.

Webster, D.J. & N.A. Streten, 1972. Aspects of Late Quaternary climate in tropical Australia. Pages 144–181 in Walker, 1972.

Whitehouse, F.W., 1940. The climates of Queensland since Miocene times. Studies in Late Geol. Hist. of Queensland. Univ. of Queensland, Dept. Geol. Publ. 2: 67–74.

Whitehouse, F.W., 1963. The sandhills of Queensland-coast and desert. Queensland Naturalist, 17, 1/2: 1–10.

Zeuner, F.E., 1942. Geology, climate and faunal distribution in the Malay Archipelago. Proc. R. ent. Soc., London, A 16: 117–123.

Zeuner, F.E., 1943. The biogeographic division of the Indo-Australian Archipelago. Proc. Linnean Soc., London, Session 154: pp 157–165.

Zwierzycki, J., 1920. Zijn de Indische petroleumterreinen in het bijzonder die op Sumatra peneplains of abrasievlakken. De Mijningenieur, 1: 3–5.

D.A. HOOIJER
Rijksmuseum van Natuurlijke Historie
Leiden

QUATERNARY MAMMALS WEST AND EAST OF WALLACE'S LINE

Wallace's Line marks the edge of the continental shelf in Indonesia: it separates Borneo from Celebes, and Bali from Lombok. At the height of Pleistocene glaciations the shelf area, Sundaland, was dry or mostly so, and land mammals could make their way from Southeast Asia to what are now the islands of Sumatra, Java, and Borneo. Macassar Strait between Borneo and Celebes is so wide and deep that it acted as an effective barrier in the dispersal of a number of mammalian species of Asiatic origin, but the Lesser Sunda Islands chain extending east from Bali was more easily accessible: some of the water gaps now separating these islands dried up in the Pleistocene, and Indo-Malayan species penetrated to Flores and Timor. On the other hand, Australian elements coming in from the east had to jump the water gaps between Australia and New Guinea (Sahulland) to Timor and the islands in the Banda Sea, which were considerable even in times of glacially lowered sealevels. Cuscuses (genus Phalanger) got to Timor, and spread through the Moluccas, establishing two endemic species in Celebes. They never crossed Wallace's Line. As a whole, the recent fauna of Celebes is three-fourth Asiatic in origin (Mayr, 1945).

If we now turn from the present to the past, and look into the fossil record, we find that many of the extant species of the Indo-Malayan area already existed in the Pleistocene, and that the fauna of Sundaland was more uniformly distributed over Java, Sumatra, and Borneo (Table 1). As far as the ten well-known, terrestrial Asiatic mammals enumerated there are concerned, Java has less than half of them today, but Sundaland species now extinct in Java do occur as fossil remains in the Djetis and Trinil faunas of that island, post-Villafranchian assemblages made world-renowned by Pithecanthropus, Java Man, first found at Trinil by Dubois in 1891/92. In 1890 he already unearthed a juvenile mandibular fragment of a hominid at Kedungbrubus in what we now know is the Djetis fauna, in beds called the Putjangan Beds underlying the Kabuh Beds with the Trinil fauna, both exposed in the area of the Sangiran dome where Von Koenigswald made further discoveries of fossil hominids between 1936 and 1941. The Djetis fauna has more extinct species than has the Trinil fauna, of which I will mention a few. First of all a species originally described from

the Pinjor horizon of the Upper Siwaliks of India as Elephas planifrons
Falconer & Cautley, which survives in the post-Villafranchian Djetis beds
just as it does in the Lang Son fauna of Indo China, where it is associated
with more advanced elephants like Elephas namadicus Falconer & Cautley
(Hooijer, 1955: 94, 95: Karang Jati, Tritik; Patte, 1955). The earliest
occurrence of Elephas planifrons in Java is at Tjijulang in the western part
of the island (Von Koenigswald, 1951; Hooijer, 1953c: 225, 227), where
it occurs in a fully extinct faunal assemblage antecedent to the Djetis fauna
and generally considered to represent a Villafranchian stage.

The Villafranchian is a type of fauna we now know to have ranged from
about 4 to about 2 million years ago: the earliest Villafranchian faunas such
as those of Vialette and Etouaires are dated at 3.8 and 3.4 million years,
and the latest, at Chilhac and Coupet, at 1.9 to 1.8 million years (Van
Couvering, 1972: 257). If, with Van Couvering (1972: 252) and Maglio
(1973: 68), we place the Plio/Pleistocene boundary at roughly 2 million
years, then Mount Coupet emerges as earliest Pleistocene, and older
Villafranchian faunas are Pliocene. For long years Elephas planifrons was
held to be the most primitive species of true elephant and was believed to
have inhabited Europe and Africa as well as Asia. Its sudden appearance
in European and Asiatic deposits was taken as a marker for the beginning
of the Villafranchian and was used for faunal correlation (eg, Hooijer,
1956). We know better now, thanks to the epoch-making studies of Maglio
(1970, 1973). The true elephants originated in Africa, several million
years before their emergence in Europe and Asia. The African origin of
the elephants was expounded by Osborn (1934), who held Archidiskodon
subplanifrons Osborn (including A. proplanifrons Osborn) from South
Africa to be indubitably ancestral to Archidiskodon planifrons. As pointed
out by Maglio, subplanifrons is on the Mammuthus lineage, and planifrons
belongs to Elephas.

Elephas planifrons, then, is a strictly Asiatic species, derived from
Primelephas gomphotheroides Maglio of Africa via Elephas ekorensis Mag-
lio at the 4 million year level in Kenya (see Fig.1, taken from Maglio,
1973, Fig.19), with occurrences of primitive stages along the expansion
route from Africa such as the Bethlehem elephant, the earliest stage of
Elephas recorded outside Africa (Hooijer, 1958a; Maglio, 1973: 44–45).
In the Siwaliks it occurs first in the Tatrot horizon, and in Java in the
Tjijulang Beds.

In the Djetis Beds, but also in the area of the Sangiran dome, there is
a pygmy elephant originally named Stegodon hypsilophus Hooijer but which,
in concurrence with Maglio (1973: 46–47), I now regard as belonging to
Elephas celebensis (Hooijer, 1974b). It is less advanced in the process of
size reduction and relative crown heightening than the pygmy elephant from
Celebes (of which more later), but like the Celebes elephant was derived
from Elephas planifrons. It is probable that isolation somehow was involved in its coming into being right there in the middle of the island of Java,
whence we know its mother species Elephas planifrons already from the
Tjijulang Beds.

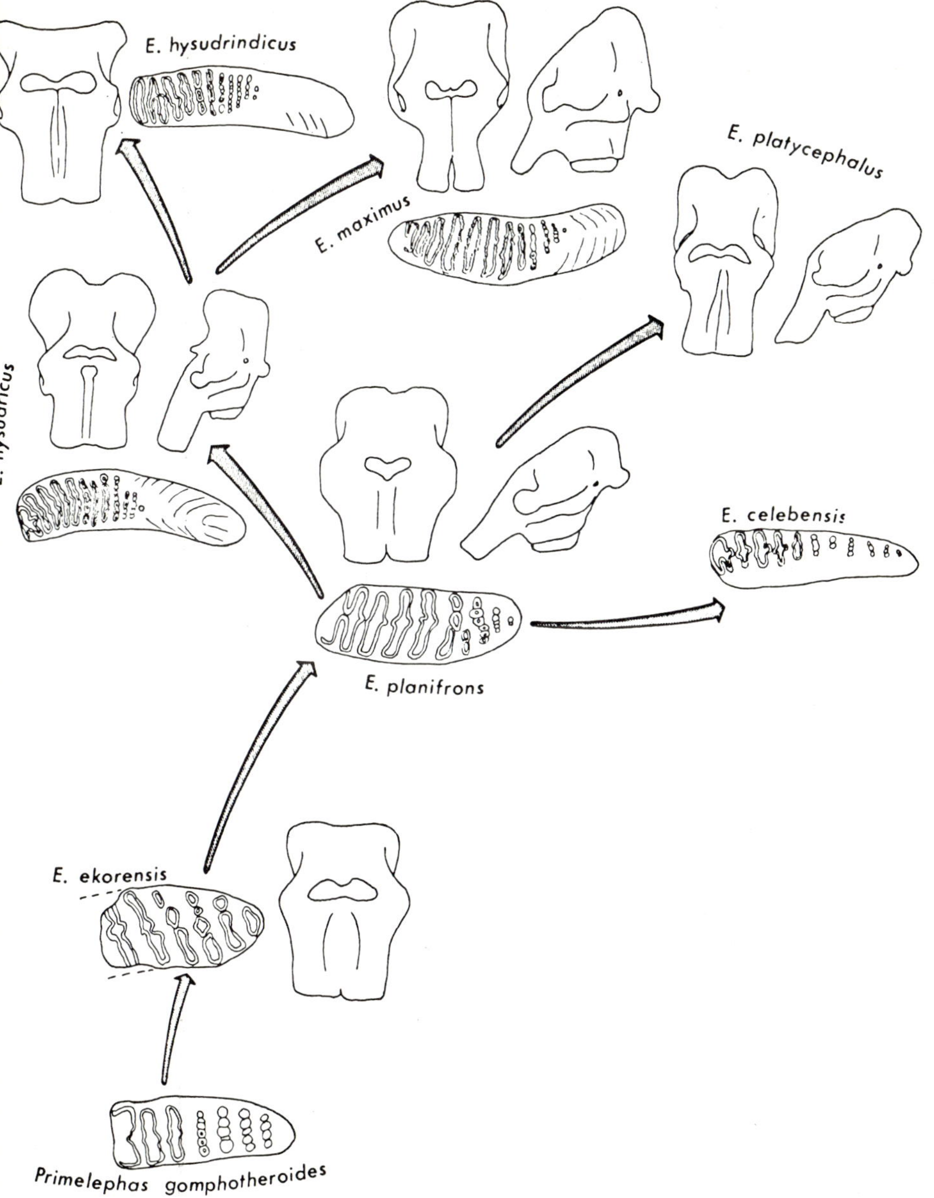

Figure 1. Diagrammatic representation of the phyletic relationships between seven species of the Asiatic branch of the Elephas complex and Primelephas gomphotheroides (after Maglio, 1973, Fig. 19).

Figure 2. Weaiwe, near Atambua, Timor. Fossils said to come from coarse gravel layer in centre foreground. June, 1970. I.C. Glover phot.

The elephant of the Trinil Beds is not Elephas planifrons but a much more advanced species, Elephas hysudrindicus Dubois (see Hooijer, 1955: 110-131), which also figures in the diagrammatic representation taken from Maglio (Fig.1). It is not transitional between Elephas hysudricus Falconer & Cautley and Elephas indicus G. Cuvier (recte: Elephas maximus L.) as Dubois thought it was, but could have been derived from an earlier stage of the Elephas hysudricus specific lineage. Elephas hysudricus and Elephas planifrons are both present in the Pinjor horizon of the Siwaliks, and it seems probable that both had a common ancestry in the ekorensis complex (Maglio, 1973: 83-86).

The Djetis and Trinil faunas contain an advanced leptobovine, Epileptobos groeneveldtii Dubois, quite abundant in the Dubois Collection, more progressive than Leptobos of the Villafranchian of Eurasia in having larger and more backwardly placed horn cores, in both sexes instead of only in the males as in Leptobos, in the interparietal being fully deflected into the occipital plane (as in Bibos), and in being more hypsodont. Von Koenigswald denied its presence in the Trinil fauna, and named the Djetis form Leptobos cosijni, but we have a specimen from Trinil, where Djetis Beds are absent, and the skull of Leptobos cosijni has all the specific characters of that of Epileptobos groeneveldtii (Hooijer, 1958b). I mention this especially because Leptobos is a characteristically Villafranchian bovid that has been much used in correlation of faunas, as has "Archidiskodon".

It should be emphasized that even today we do not have very precise radiometric datings of the Trinil or the Djetis Beds: there is a date for a tuff several meters below the Homo erectus child skull site in the Djetis beds at Modjokerto of 1.9 ± 0.4 million years (Jacob, 1972), indicating that this is from about half a million years older to half a million years younger than the upper limit of the Villafranchian, and there is a date for tektites averaging 600,000 years, but whether this has any valid bearing on the age of the Trinil beds is open to question. In a correlation by Maglio (1973: 71) Trinil is placed just above the 1 million year level, Djetis in the 1 to 2 million year interval, and Tjijulang in the 2 to 3 million year interval.

Turning now to Sumatra: no Pleistocene mammal bearing beds have so far been found in this island, and the records of the panther and the banteng from Sumatra are based on subfossil remains found in the caves of the Padang Highlands explored by Dubois (see Table 1).

The Bornean records of the tiger and the tapir are from Niah Caves, Sarawak. It is from these caves, at a level dated at 40,000 years, that we have the extinct giant pangolin, Manis palaeojavanica Dubois. The fossil sites in Java from which remains of this species have been recorded (Kedungbrubus, Tjitarum Valley, and Gunungbutak) may all be from the Djetis deposits: the Niah Caves record shows that the species survived well into the Late Pleistocene (for references, see Hooijer, 1974a).

The Indo-Malayan species listed in Table 1 were stopped down in their dispersal by Macassar Strait: none of them has ever been found in Celebes

Table 1.   Recent (x) and extinct (o) mammals of Sundaland

	Java	Sumatra	Borneo
Orang-utan	o	x	x
Siamang	o	x	
Tiger	x	x	o
Panther	x	o	
Malay Bear	o	x	x
Elephant	o	x	x
Malay Tapir	o	x	o
Java Rhinoceros	x	x	
Sumatran Rhinoceros		x	x
Banteng	x	o	x

either living or fossil. In 1947 there began a series of discoveries of fossil vertebrates in southwestern Celebes by Van Heekeren, which were
beyond anything zoogeographers would have believed to exist. Somehow, in
the Early Pleistocene, Elephas planifrons crossed the water gap from
Sundaland to Celebes, and the immigrants underwent a very rapid reduction in size that resulted in the pygmy Elephas celebensis (Hooijer, 1949).
No remains of the larger Elephas planifrons, from which the pygmy must
needs be derived as there is no other known suitable ancestor, have so far
been recovered in Celebes. When the overseas colonization took place we
have no means of knowing, nor do we know by which route. Macassar
Strait can be overcome by island hopping round about it: via Palawan to
the Philippines and then down south via the Sangihe Islands to the northern
peninsula of Celebes. In Mindanao and Luzon fossil mammals of Asiatic
origin have been found: this is a possible migration route. The pygmy
elephant of Celebes is just a fifty per cent scale reduction of Elephas
planifrons in every way; this struck me right from the beginning. Entire
lower molars are half as long and wide as their homologues in Elephas
planifrons (Hooijer, 1953a); both species have functional premolars
(Hooijer, 1953c), and the occasional presence of functional tusks in the
mandible (Hooijer, 1954b), which puzzled me at first as these structures
were not then known to exist in Elephas planifrons, has since been
explained as a result of dwarfing: Maglio (1973: 46) found vestigial incis-
ive chambers in several mandibles of Elephas planifrons unrecorded
before, similar to those in species possessing formed mandibular tusks
such as Primelephas gomphotheroides. They simply redeveloped in the
island form, possibly through paedomorphosis (Maglio, 1970: 10; 1973: 46).

The stratigraphy of the pygmy elephant-bearing deposits in Celebes is
as yet not clear, but so much is clear that the most complete specimens,
such as a good skull portion with the entire palate and the bases of the
tusks, come from the highest terrace along the Walanae River in the
Tjabenge area (Hooijer, 1972b, Pl.1). As the all-important fronto-
parietal surface is not preserved, the skull has not been illustrated in Fig.
1; as far as the skull affords means of comparison with that of Elephas
planifrons the tusks seem to be directed a little more downward in Elephas

celebensis than they are in that species, and they are certainly more
closely approximated.

As to the origin of another element in the fossil Celebes fauna we are
less secure: this is a precinctive suid species called Celebochoerus
heekereni Hooijer (1954a). It is most abundant at the Tjabenge sites,
although the skull is as yet unsatisfactorily known. Celebochoerus males
had huge tusks in the upper jaw; in the females they were smaller. The
premolars are potamochoeroid, and the molars relatively simple in
structure. I regard Celebochoerus as most closely related to the Early
and Middle Siwalik genus Propotamochoerus, derived from the same ancestral stock, and evidently with a long and independent evolutionary history
behind it. There is no evidence of any real near relationship with either
of the two forms of suids living in Celebes today, the babirusa and Sus
celebensis Müller & Schlegel (which incidentally was recently found as a
fossil in the same area: Hooijer, 1969b). In recent papers, Thenius (1970,
1972) has dealt with Celebochoerus, and he placed it on one of the lineages
of the Suinae dating back to the Miocene, parallel to the Propotamochoerus-
Potamochoerus line, a product of the secondary radiation of the Suidae in
the Late Tertiary which led to the Suinae as well as to the Phacochoerinae.
The presence of three genera of suids in Celebes (Celebochoerus, Babyrousa, and Sus), which must have split off from the main stock of the
Suidae at different times in the Tertiary (the babirusa is traced back to
the Oligocene, again with no known forms on the postulated lineage but the
end product Babyrousa), is indicative of the several invasions of ancestral
suid types to Celebes over a great length of time, but as is the case with
Elephas celebensis the ancestral forms have not been found.

We are luckier with another pygmy proboscidean that occurs in the fossil state in Celebes, Stegodon sompoensis Hooijer (1964a). This is a fifty
per cent scale reduction of Stegodon trigonocephalus Martin from Java.
Stegodonts are not elephants; they have long been considered ancestral to
them but constitute a group all by themselves: the Stegodontidae. They
seem to represent a specialized line of mammutid descent, convergent on
the elephant-type of masticatory pattern (Maglio, 1973: 16). They are
quite abundant as fossils in Asia, and the species known from the best
materials is Stegodon trigonocephalus from Java (Hooijer, 1955: 17–86).
It is an advanced species of Stegodon, slightly smaller than Stegodon insignis Falconer & Cautley from the Upper Siwaliks, and ranges in Java from
the Late Pliocene to the Late Pleistocene. It has not been found in Sumatra
or Borneo, but molar fragments indistinguishable from it are known from
Mindanao and Luzon (Naumann, 1890; Beyer, 1949; Von Koenigswald, 1956).

The first Stegodon molars recorded from Celebes (Hooijer, 1953b)
were incomplete, of uncertain serial position, and I could not make up my
mind as to whether they represented a pygmy form or one of continental
size. The name-bearer of Stegodon sompoensis (Hooijer, 1964a) is an
unmistakable dwarf milk molar, and further material obtained during our
expedition to Celebes in 1970 proved that there were actually two distinct
species of Stegodon in the Pleistocene of Celebes, a large one, of the same

size and with the same low molar crowns as Stegodon trigonocephalus from
Java, and a fifty per cent scale reduction of the latter with, however, relatively higher molar crowns. This is exactly the same association that we
know to have existed in the Philippines: on Mindanao there is Stegodon cf
trigonocephalus, and a smaller form that is higher-crowned, Stegodon
mindanensis (Hooijer, 1972b). When the new material from Celebes was
described I had already seen and published on the Flores and Timor forms,
both discoveries by Dr Th. Verhoeven.

The Flores stegodont, Stegodon trigonocephalus florensis Hooijer
(1957), is somewhat smaller, on an average, and is more hypsodont than
the Java Stegodon trigonocephalus, and it is quite evident that it made its
way to Flores past the Lesser Sunda Islands from Java, at times of glacially lowered sealevel. We have no radiometric datings on the deposits in
Flores from which these specimens came, and it is hoped that these may
become available in the future. Along with this form there is a pygmy
stegodont in Flores also: I described a couple of milk molars from that
island obtained by Mr Hartono of the Geological Survey of Indonesia that
are half as large as their homologues in Stegodon trigonocephalus but
refrained from naming them (Hooijer, 1964b), noting that the Flores
pygmy stegodont could not be differentiated from that of Celebes on the
meagre material available. In the same year, Verhoeven (1964) announ-
ced the discovery of Stegodon fossils from Timor, and the first found
specimen, an incomplete molar, was made the type of Stegodon timorensis Sartono (1969). Specimens later found in the same area near Atambua
in Indonesian Timor, and doubtless specifically identical with that described
by Professor Sartono, proved that this is a pygmy form half as large in
linear dimensions as Stegodon trigonocephalus, but that there is also a
larger form, less well represented, indistinguishable from Stegodon
trigonocephalus (Hooijer, 1969a). Thus, the two islands, Flores and
Timor, appeared to have the same association of a large and a small
species of Stegodon, pointing, it seemed, to two separate invasions unless
the larger form is the ancestor of the smaller, as may seem most likely
as the larger form already marks a step from Stegodon trigonocephalus of
Java in the direction of the pygmy in its slightly reduced size and relative
heightening of the molars. We do not know whether the two forms are
strictly contemporaneous since also in Timor we lack stratigraphic control
on the specimens and radiometric dates are not available as yet. However,
the resemblance between the stegodonts of Flores and those of Timor is so
close that parallel independent evolutionary change to the pygmy condition
now seemed less likely than to postulate that migrations back and forth
were possible at the time these forms were evolving in Flores and Timor
(Hooijer, 1972a). In order for this to be possible, the sea gaps between
Flores and Timor must have been less formidable than they are now, quite
beyond the power of an elephant or a Stegodon to swim across.

A study by Dr M.G. Audley-Charles has eliminated the difficulties in
assuming inter-island migrations of stegodonts in the Pleistocene: on the
basis of evidence from the Viqueque Formation in Portuguese Timor a land

connection between Timor and Flores via Alor may be postulated sometime
in the Early and Middle Pleistocene, after which, in the Late Pleistocene
and Holocene, downfaulting occurred involving at least 3000 meters. The
deeply incised alluvial terraces from which the stegodont bones have been
obtained also indicate this recent uplift. And the formidable Early and
Middle Pleistocene uplift in the Flores-Timor region may have extended
northward, and affected the now nearly 3000 meters deep submarine ridge
between southwestern Celebes and northeastern Flores, so that the stegodonts could wander back and forth between these islands also. This ridge
system, Salajar-Matjan-Kaloatoa, may also have been elevated above sea-
level by a regional uplift that followed the plate collision between eastern
and western Celebes in the Late Pliocene-Early Pleistocene (Audley-
Charles & Hooijer, 1973). Since the stegodonts of Celebes and Flores-
Timor are so exceedingly alike (the entire molars of Stegodon sompoensis obtained in 1970 fall within the variation limits of those of Stegodon
timorensis) and there is geological evidence now for an island connection,
it seems unnecessary to hold on to the view that the stegodonts of Celebes
and Flores-Timor evolved in geographic isolation, which I still adhered to
in 1972 (Hooijer, 1972b).

Consequently, it seems that one and the same species, Stegodon sompoensis, the smallest and highest-crowned Stegodon ever, inhabited
Celebes, Flores and Timor in the Pleistocene: it was just one interbreed-
ing population. Whether the Mindanao pygmy stegodont is specifically the
same as Stegodon sompoensis is only to be established on better material
from the Philippines. For the present we may consider Celebes, Flores
and Timor to have been one, the homeland of the pygmy stegodonts, which
may be called Stegoland (as suggested to me by Mr T. Harrisson). Situated between Sundaland and Sahulland, Stegoland is a zoogeographic sub-
unit of the area intermediate between the Indo-Malayan and Papuan-Aus-
tralian faunal regions. One of the unsolved problems of palaeozoogeography
is why other elements of the Celebes fauna such as Elephas celebensis or
Celebochoerus did not avail themselves of the opportunity of getting to
Flores and Timor. Perhaps they were less capable migrants, or the road
to the south was blocked at the time they lived on Celebes: although Elephas,
Celebochoerus and Stegodon are found together we do not have sufficient
stratigraphic evidence to prove their contemporaneity. It should also be
stated that there is little faunal likeness between Celebes on the one hand,
and Flores-Timor on the other as far as the living animals are concerned:
recent zoogeographers like Stresemann (1939: 339) and Mayr (1944: 178)
have discarded the Flores-Celebes land bridge that was once postulated by
the Sarasins to account for the present-day Celebes fauna. The cave rats
of Timor and Flores, Coryphomys and a couple more new genera to be described from Timor, dating back to 13,000 years BC (Dr Ian Glover, personal communication), and Papagomys and Spelaeomys from Flores, show
no close resemblance and, therefore, there was no faunal interchange
between these islands at that time.

In all of the three islands of Stegoland we have evidence of continental-

sized stegodonts beside the pygmy forms, and it is probable that the former represent the immigrant colony that gave rise to the insular dwarfed
forms at a later date. Much stratigraphic work remains to be done in these
three islands to settle the temporal relationships between the stegodonts,
which we now have in two size classes only, with no intermediates. Celebes,
Flores and Timor form a most promising area for further geological and
palaeontological research in Indonesia, no less than Java on which most
attention has been focused because of the fossil hominids there. Palaeolithic
tools are known from all the islands, and the incentive to go there in search
of Early Man may also be beneficial for the study of the non-hominid fossils
in Stegoland.

REFERENCES

Audley-Charles, M.G. & D.A. Hooijer, 1973. Relation of Pleistocene
    migrations of pygmy stegodonts to island are tectonics in eastern Indo-
    nesia. Nature, 241 (5386): 197-198.
Beyer, O., 1949. Outline review of Philippine archaeology by islands and
    provinces. Philipp. J. Sci., 77: 205-374.
Couvering, J.A. van, 1972. Radiometric calibration of the European
    Neogene, pages 247-271, in W.W. Bishop & J.A. Miller (eds), Calib-
    ration of Hominoid Evolution. New York, Scottish Academic Press.
Hooijer, D.A., 1949. Pleistocene Vertebrates from Celebes. IV. Archidis-
    kodon celebensis nov. spec. Zool. Meded. Leiden, 30 (14): 205-226.
Hooijer, D.A., 1953a. Pleistocene Vertebrates from Celebes. V. Lower
    molars of Archidiskodon celebensis Hooijer. Ibid, 31 (28), 311-318.
Hooijer, D.A., 1953b. Pleistocene Vertebrates from Celebes. VI. Steg-
    odon spec. Ibid, 32 (11): 107-112.
Hooijer, D.A., 1953c. Pleistocene Vertebrates from Celebes. VII. Milk
    molars and premolars of Archidiskodon celebensis Hooijer. Ibid, 32
    (20): 221-231.
Hooijer, D.A., 1954a. Pleistocene Vertebrates from Celebes. VIII. Den-
    tition and skeleton of Celebochoerus heekereni Hooijer. Zool. Verh.
    Leiden, 24: 1-46.
Hooijer, D.A., 1954b. Pleistocene Vertebrates from Celebes. XI. Molars
    and a tusked mandible of Archidiskodon celebensis Hooijer. Zool.
    Meded. Leiden, 33 (15): 103-120.
Hooijer, D.A., 1955. Fossil Proboscidea from the Malay Archipelago and
    the Punjab. Zool. Verh. Leiden, 28: 1-146.
Hooijer, D.A., 1956. Archidiskodon planifrons (Falconer & Cautley) from
    the Tatrot zone of the Upper Siwaliks. Leidse Geol. Med., 20: 110-119.
Hooijer, D.A., 1957. A Stegodon from Flores. Treubia, 24: 119-129.
Hooijer, D.A., 1958a. An Early Pleistocene mammalian fauna from Beth-
    lehem. Bull. Br. Mus. (Nat. Hist.), Geol., 3 (8): 265-292.
Hooijer, D.A., 1958b. Fossil Bovidae from the Malay Archipelago and the
    Punjab. Zool. Verh. Leiden, 38: 1-112.

Hooijer, D.A., 1964a. Pleistocene Vertebrates from Celebes. XII. Notes on pygmy stegodonts. Zool. Meded. Leiden, 40 (7): 37–44.

Hooijer, D.A., 1964b. On two milk molars of a pygmy stegodont from Ola Bula, Flores. Bull. Geol. Surv. Indonesia, 1 (2): 49–52.

Hooijer, D.A., 1969a. The Stegodon from Timor. Proc. kon. Ned. Akad. Wet. Amsterdam, B 72: 203–210.

Hooijer, D.A., 1969b. Pleistocene Vertebrates from Celebes. XIII. Sus celebensis Müller & Schlegel, 1845. Beaufortia, 16 (222): 215–218.

Hooijer, D.A., 1972a. Stegodon trigonocephalus florensis Hooijer and Stegodon timorensis Sartono from the Pleistocene of Flores and Timor. Proc. kon. Ned. Akad. Wet. Amsterdam, B 75: 12–33.

Hooijer, D.A., 1972b. Pleistocene Vertebrates from Celebes. XIV. Additions to the Archidiskodon-Celebochoerus fauna. Zool. Meded. Leiden, 46 (1): 1–26.

Hooijer, D.A., 1974a. Manis palaeojavanica Dubois from the Pleistocene of Gunung Butak, Java. Proc. kon. Ned. Akad. Wet. Amsterdam, B 77: 198–200.

Hooijer, D.A., 1974b. Elephas celebensis (Hooijer) from the Pleistocene of Java. Zool. Meded. Leiden, 48 (11): 85–93.

Jacob, T., 1972. The absolute date of the Djetis beds at Modjokerto. Antiquity, 46 (182): 148.

Koenigswald, G.H.R. von, 1951. Ein Elephant der planifrons-Gruppe aus dem Pliocaen West-Javas. Eclogae geol. Helv., 43: 268–274.

Koenigswald, G.H.R. von, 1956. Fossil mammals from the Philippines. Proc. fourth Far-Eastern Prehistory Congress, Quezon City, 1: 339–362.

Maglio, V.J., 1970. Four new species of Elephantidae from the Plio-Pleistocene of northwestern Kenya. Breviora, Mus. Comp. Zool., 341: 1–43.

Maglio, V.J., 1973. Origin and evolution of the Elephantidae. Trans. Amer. Phil. Soc. Philadelphia, ns, 63 (3): 1–149.

Mayr, E., 1944. The birds of Timor and Sumba. Bull. Amer. Mus. nat. Hist., 83: 123–194.

Mayr, E., 1945. Wallace's Line in the light of recent zoogeographic studies. Pages 241–250 in P. Honig & F. Verdoorn (eds), Science and Scientists in the Netherlands Indies. New York, Board for the Neth. Indies, Surinam and Curacao.

Naumann, E., 1890. Stegodon mindanensis, eine neue Art von Uebergangs-Mastodonten. Zeitschr. Deut. geol. Ges., 42: 166–169.

Osborn, H.F., 1934. Primitive Archidiskodon and Palaeoloxodon of South Africa. Amer. Mus. Novit., New York, 741: 1–15.

Patte, E., 1955. Un éléphant Archidiskodon des Brèches de Langson (Vietnam). Bull. Soc. géol. France, (6), 4: 505–508.

Sartono, S., 1969. Stegodon timorensis: a pygmy species from Timor (Indonesia). Proc. kon. Ned. Akad. Wet. Amsterdam, B 72: 192–202.

Stresemann, E., 1939. Die Vögel von Celebes. I. Die ornithologische Erforschung von Celebes. II. Zoogeographie. J. Ornith., 87: 299–425.

Thenius, E., Zur Evolution und Verbreitungsgeschichte der Suidae (Artiodactyla, Mammalia). Z. Säugetierk., 35 (6): 321–342.

Thenius, E., Microstonyx antiquus aus dem Alt-Pliozän Mittel-Europas. Zur Taxonomie und Evolution der Suidae (Mammalia). Ann. Naturh. Mus. Wien, 76: 539–586.

Verhoeven, Th., 1964. Stegodon-Fossilien auf der Insel Timor. Anthropos, 59: 634.

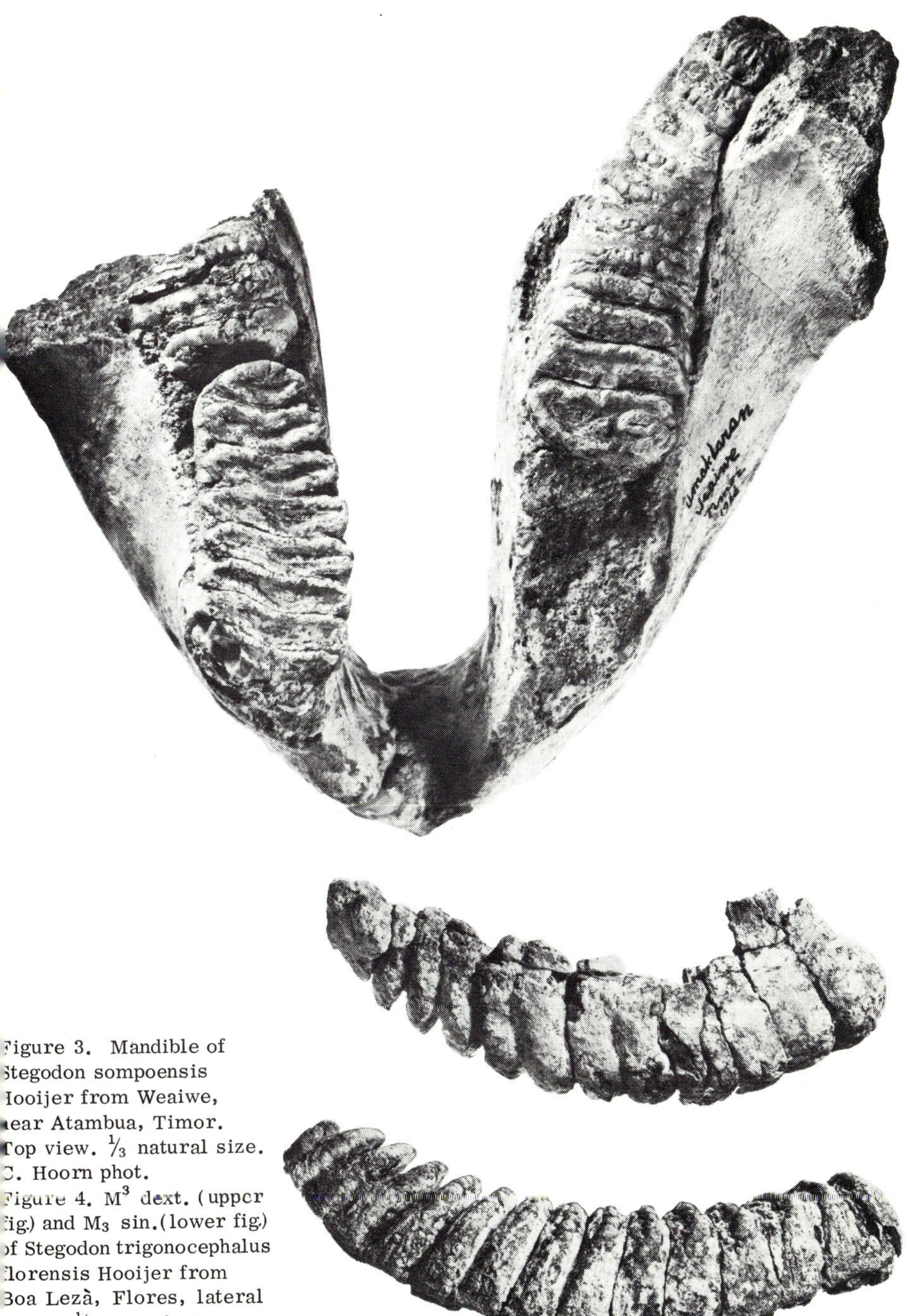

Figure 3. Mandible of
Stegodon sompoensis
Hooijer from Weaiwe,
near Atambua, Timor.
Top view. ⅓ natural size.
C. Hoorn phot.
Figure 4. M³ dext. (upper
fig.) and M₃ sin.(lower fig.)
of Stegodon trigonocephalus
florensis Hooijer from
Boa Lezà, Flores, lateral
views. ¼ natural size.
C. Hoorn phot.

Figure 5. Third lower milk molars of Stegodon trigonocephalus Martin (Java), Stegodon trigonocephalus florensis Hooijer (Flores), and Stegodon sompoensis Hooijer (Flores). Crown views. $\frac{3}{4}$ natural size. C. Hoorn phot.
Third upper milk molars of Stegodon trigonocephalus Martin (Java), and Stegodon trigonocephalus florensis Hooijer (Timor). Crown views. $\frac{6}{7}$ natural size. C. Hoorn phot.

Figure 6. M_1 dext. of Elephas celebensis (Hooijer) from Sompoh, Celebes. Crown (upper) and lateral (lower) views. Natural size. H.F. Roman phot.
Figure 7. Skull of Elephas celebensis (Hooijer) from Tjabkangè, Celebes. Left lateral view. $\frac{1}{3}$ natural size. T. Asmar phot.

Figure 6.

Figure 7.

Figure 8. Palate of Celebochoerus heekereni Hooijer from Marale, Celebes.
Palatal view. $^4/_5$ natural size. C. Hoorn phot.

Figure 9. A giant among the extinct mammals West of Wallace's Line: Manis palaeojavanica Dubois. Total length 2.5 m. W.C.G.Gertenaar del.

H.R. VAN HEEKEREN

CHRONOLOGY OF THE INDONESIAN PREHISTORY

CHRONOLOGY OF THE INDONESIAN PREHISTORY

PLEISTOCENE

GEOLOGY	FAUNA	HOMINIDS
LATE UPPER Deepest cave deposits	Recent	Homo sapiens: Niah cave man
EARLY UPPER Notopuro beds	? ?	? ?
Ngandong twenty m. river-terraces	Ngandong fossil	Homo soloensis
Baksoka twenty m. river-terraces	? ?	? ?
Mengeruda river deposits	Mengeruda fossil	? ?
Tjabengè tectonic 50 m.	Tjabengè fossil	? ?
Sangiran gravel sheet	? ?	
MIDDLE Kabuh beds: conglomerates, tuff and sandstone	Trinil fossil	Pithecanthropus erectus
LATE LOWER Putjangan beds; black clays; lower volcanics	Djetis fossil	Pithecanthropus robustus Meganthropus palaeojavanicus
EARLY LOWER	Kali Glagah Tji Djulang (Villafranchien)	? ?

ARCHAEOLOGY	ECONOMY	DATING
Non-ceramic pebble- and flake industries; cave burials; red ochre	Nomadic food gathering	30 - 50,000
Sangiran flake industry	Nomadic food gathering and hunting	? ?
Ngandong flake and bone industry		est. 100 - 200,000
Flake-cum-Pebble tool industry. (Patjitan)		
Flake-cum-Pebble tool industry. (Timor, Flores)		
Flake industry (Tjabengè, Sulawesi)		
Flake industry?		av. 830,000
? ?	? ?	av. 1,900,000

N.B. There may be great difference in time and a
long survival of old forms in some areas

GEOLOGY	FAUNA	HOMINIDS
Tropical soils	Domesticated animals: pig, horse, dog, fowl	Palaeo-mongoloids
Tropical soils	Domesticated animals: pig, dog, fowl	Palaeo-mongoloids
Tropical soils	Recent	Palaeo-mongoloids Oceanic negroids

GEOLOGY	FAUNA	HOMINIDS

ARCHAEOLOGY	ECONOMY	DATING
		B.P.
LATE BRONZE-IRON AGE Majority iron tools. Minority bronze tools. Dolmen, stone vats, stone cists; menhirs and menhir statuettes; glass beads, simple globular pottery.	Wet-rice cultivation; trade-routes by land and by sea; sailing boats; head-hunting; class stratification; animal sacrifice; feast-of-merit; ancestor cult.	1000 – 2000
EARLY BRONZE-IRON AGE Majority bronze tools and ornaments. Minority iron tools. Stone sarcophagi with human, animal and geometric designs; coastal settlements and necropols; metal kettle-drums Heger I type; socketed axes and lance-heads; gold objects; bronze figurines; daggers with bronze and iron blades; a great variety of pottery; geometric decor-ations; spirals, cord-marked, whirl motifs, meanders, circlets joined by oblique tangents, hatched triangles. Agitated animal art and deer procession. Pedestal bowls. Glass and carnelian beads. Textile.	Wet-rice cultivation; class-stratification with hereditary chieftains; human and animal sacrifice; feast-of-merit; ancestor cult; kettle-drum trade route; sailing boats with and without outriggers.	2000 – 3000
LATE-NEOLITHIC Rectangular polished stone adzes. Sa-huynh pottery; bow-and-arrow of wood; First textile. Paddle and anvil pottery manufacture.	Oldest wet-rice cultivation. Small villages with chieftains; head-hunting; canoes with outriggers and rafts; tubers, millet and rice.	3000 – 4000
EARLY-NEOLITHIC Round axes; Rectangular adzes; cord-impressed pottery; incised and impressed pottery; bow-and-arrows of wood; bark-cloth	Shifting cultivation; slash-and-burn	4000 – 5000
EPI-PALAEOLITHIC Microlithic industry; kitchen-middens in caves; burial outside caves; negative hand-stencils and rock-paintings on cave walls.	Semi-nomadic food-gathering and small game hunting; shell-fish and fishing; division of labour based on sex.	5000 – 7000
Pebble-flake and shell tools; cord-marked pottery; kitchen-middens along the coast. Flexed burials in caves and kitchenmiddens; bone tools; ornaments of shell and bone and animal teeth. Red and yellow ochre.	Semi-nomadic food-gathering and small game hunting; first domesticated plants??? Division of labour based on sex	7000 – 10,000

TOM HARRISSON
University of Sussex
Brighton

4

TAMPAN: MALAYSIA'S PALAEOLITHIC RECONSIDERED

1 DISCOVERY AND EXCAVATIONS AT KOTA TAMPAN, PERAK (1936-54)

Kota Tampan, a river terrace open-site on the true right bank of the middle reaches on the Perak River, lies south of Lenggong and just north of (artificial) Lake Chenderoh in Perak (West Malaysia). It was discovered archaeologically by H.D. Collings, a curator of the then Raffles Museum, Singapore, in 1938. He brought 208 stone artefacts back to. Singapore, where I often examined them with him, later in the forties. He only published a short note (Collings, 1938), partly because he seemed uncertain exactly how far — and especially how far back — to go in interpreting his results. However, Dr Hallam Movius (1948:403) lent the great weight of his authority when he accepted Tampan's general importance and published five tool photographs. Others followed suit, without further reference to the original material.

Then, in 1954 Ann de G. Sieveking, talented archaeologist-wife of the then curator of the Taiping Museum, Perak, re-opened and extended Collings' earlier work in the immediate vicinity of original discovery. She brought back 254 stone tools to Taiping, now transferred to the Muzium Negara, Kuala Lumpur. She wrote a full, lucid account (Sieveking, 1960, and principally 1962), in association with an Australian Army geologist, D.C. Walker (1956, 1962) on secondment to the Geological Survey in Malaya. His interpretations were fundamental to Mrs Sieveking's argument, especially on the subject of Pleistocene sea-levels and the relationship of the Tampan gravels thereto. Collings had no close geological support — and felt the loss.

Both the Collings and Sieveking results came from a single small acre of cleared land under rubber or (round a ruined house) scrub. Apart from the stone tools nothing else was found to indicate human activity. No fossils or other associated materials were obtained. Nor have the "Tampanian" sorts of artefact (see Sections 3 and 4 below) been excavated elsewhere, though Collings found a few more casually at other localities along the Perak River, as far up as Grik (Gerik), 50 km north of Lenggong (Tweedie 1953:10).

On the basis of Collings' 1938 note and Movius' 1948 extension, Tampan
soon became — as so often happens — accepted in the general literature.
By 1953, Michael Tweedie, in his excellent booklet on the Malayan Stone
Age, had (in private, somewhat reluctantly) to note it the only Palaeolithic site known in the peninsula. As he there put it: "references to the
Tampan or Tampanian Culture, as it has come to be called, are frequent
in subsequent literature" (1953: 9).

Mrs Sieveking's enthusiastic report of 1962 further raised its status
into the lofty isolation of the only fully Pleistocene site for Malaya. Indeed
she put it far back (much beyond the possible first date for Homo sapiens)
by concluding: "the Tampanian industry is of possible late First Interglacial
or more probably of early Second Glacial date" (1962: 111). And again:
"... this gives the Tampanian a date as early or earlier than that of any
other Palaeolithic industry in the Far East" (1962: 120). She goes on,
always relying primarily on Pleistocene sea-levels supposed for the Tampan gravel, to declare: "This new low date for a South-East Asian Palaeolithic industry suggests that the general ideas relating to the character
and diffusion of these Asian industries should be revised" and this leads
(pp.121-133) to far reaching theoretical concepts, based necessarily (in
accord with her training and only previous experience) upon European
models. She even re-assesses Olduvai Gorge in Kenya as a result of her
Tampan conclusions (p.28); and she firmly decides that "in fact, the per-
spective in South and East Asia has been changed by a new early date for
Tampan" (p.133).

Unfortunately, intensified archaeological research between Africa and
the Pacific in the twenty years since the last Tampan 'dig' have failed to
adduce any fresh evidence to support these 'early' views (cf.Sections 6
and 7). Before looking at that aspect, it will be necessary first to glance
at the Tampanian as presently understood (Section 3), then take a fresh
look at the site itself (Sections 4 and 5).

If some of the side-effects of the following exploration are critical, it
must be emphasised from the start that nothing here written is intended to
be personal; and that the writer is convinced any mistakes, if made, were
made in good faith. Yet an obligation to get at the facts inside Malaysia
cannot be put aside for personal reasons. Tampan continues to be accepted
outside at its early face value over a wide spectrum of the best authoritative
contemporary literature on prehistory — for instance, J.P. Mulvaney's
excellent Prehistory of Australia, 1969: 164, citing Sieveking's supposed
African parallels; or the new edition of H.R. van Heekeren's handy com-
pendium, The Stone Age in Indonesia, 1972: 38.

3 DEFINITIONS AND LIMITATIONS FOR "TAMPANIAN CULTURE"

Michael Tweedie, who worked all his Asian life (1932-60) in the Singapore
Museum and ended up as its director, described Tampanian tools as "so
crude and roughly finished as to be recognisable as artefacts only to an
experienced eye" (1953: 9; cf.Section 6 below).

Further definition and discussion are supplied in Mrs Sieveking's subsequent report, with especial reference to the 254 stones she considered
artefacts from her several trenches on the high river bank. She begins with
"the general appearance of the industry is of very poor quality", partly
because most of the tools ("the great majority") are made of rough-grained
quartzite, as well as some on pebbles of quartz which shatter rather than
flake (1962: 112). There are many flakes ("of a simple Clacton type")
but twice as many pebble tools proper (more in the Collings' assemblage).
In classifying all these, Sieveking based her categories "on the general
appearance and probable function of the various artefacts", although ad-
mitting difficulties with "so elementary an industry"; and, one may add,
we know absolutely nothing of local Palaeolithic life-styles and tool
usages, nor how important stone really was compared with bone, shell,
wood and bamboo (cf.Harrisson, 1974).

Nevertheless the classification adopted is elaborate, including picks
and hammerstones; large, medium and small points; unifacial and bifacial
cleavers, rough cleavers, small cleavers; heavy scrapers, large and small
scrapers; "hand axes" and undifferentiated flakes —, each with a paragraph
in small type (pp. 115-119).

Looking afresh at these stones, after nearly three decades of experience
elsewhere in the region, one is forced to wonder at the confidence which
enabled any functional distinction in western minds to cover many, even
the majority of them. Rather, one is reminded of the mass of rough
shapes, familiar to us now from much later sites in Palawan and Sabah
(East Malaysia). To this we will return further in Section 6 below.

4 THE PERAK RIVER RE-VISITED (1974)

The present writer was long puzzled by the Tampanian. These doubts were
temporarily soothed by Mrs Sieveking's powerful arguments. But since
then, own experience with the zonal Palaeolithic has revived uncertainty
over a fairly wide range of accepted attitudes on the earlier stone age, both
in the islands and on the adjacent mainland. The Conference on the Far
Asian Palaeolithic, at McGill University, Montreal in August 1973, also
showed a strong body of experienced opinion, from India through to Japan
and the Philippines, in favour of a new approach to the earlier periods:
and a general failure to confirm anything very early in human terms over
nearly all the region (Ikawa-Smith, 1974).

The writer therefore began by re-examining the generally accepted evidence for middle Pleistocene or earlier elements in the island of Borneo: tektite showers, fossil elephant teeth, "chopper tools", for instance. In no case did these stand up to the strict test of scientifically acceptable proof of early date on provenance, laboratory tests of date or any actual relationship to human artifactual or other presence (Harrisson, 1974). This study has since been extended to adjacent territories, and is continuing.

For this reason the opportunity was grasped to revisit Kota Tampan in January 1974.

It is not an easy place to find nowadays, along a small side-road and then, after the Kota Lima Estate Buildings, through rubber along a track which ends at an abandoned European-style house sometimes described as the old Estate Bungalow or as a former Customs Post, presumably in the days when this was a trade route down and across the strong river. The track goes steeply down the high bank to the sandy flats of the river bed and a ferry point served by a small motorboat running two or more times an hour (as required). The Perak River is here some 5-600 meters wide, fast flowing. Over the other side, one soon enters wild country running up to the spinal range and the Pahang border.

Over millenia, the river bed has cut deep into the bottom of the valley, two to three miles across. The valley floor is between 60 and 75 meters above mean sea level. Steep-sided ravines cut down to the present river below. Gravel usually outcrops in the ravine banks a few feet below the tops of the ridge remnants of the old valley floor, with mostly quartzite and quartz pebbles, 3 cm to 20 cm long, well rounded and water-worn. Those of quartz are highly friable. The upper surface of this gravel is just below 70 meters m.s.l. It is among these gravels and from these pebbles that both archaeologists derived their Tampanian.

A striking feature of the place is the extaordinary quantity of volcanic ash, which forms hillocks on the ridge tops, 1.5 to 6 meters across and up to 3 meters high (cf.Walker: 105). They form in effect ash dunes.

It has been generally accepted that this ash was wind-blown probably coming from the great eruption of Mt Toba, Sumatra, sometime in the Pleistocene. By projection it has naturally been thought to give a ripe antiquity to the overlaid gravels (Tweedie, 1953: 5; cf.Walker, 1962: 105). It gave, literally, a seal of some ancestry over whatever lay below, not always clearly.

More important, however, is of course the gravel-loaded terrace, 70 meters up.

Since there is no hominid or fossil association with the "tools", and since there has been no attempt to date Tampan by any of the proven laboratory methods, a great deal has to hinge on this geological correlation. In Collings' time no such information was forthcoming. Mrs Sieveking had Dr Walker's. Her report makes full use of his findings but expresses some uncertainty about the relative antiquity of the ash — which had seemed very important to Collings (see Section 5.2). She therefore emphasises rather the river terraces around the 70 meter contour line.

Sieveking sums up Walker (1962: 110-111) as concluding that the terrace gravel was deposited "under a sea, at least 230 feet (about 70 m) higher than that of the present day". Furthermore, isostatic changes of land level "can be shown to have little importance in Malaya". Her whole argument on Tampan is based on this concept. A first Interglacial date is assumed for the old alluvium, but the coarser gravel forming and carrying the tools is considered as dating to the "recession of the sea from its former high level", and thus rather later, early Second Glacial (p.111). This, in the preamble to the discussion of the tools themselves, is not solely based on Walker's attached, independent report, but supplemented by reference to both his 1956 paper and to "a letter from Dr Walker" (no date).

After a full discussion of the stones the point is restated with emphasis in the opening sentence of 14 pages of General Conclusions, thus: "In the first part of this paper it has been demonstrated that the Palaeolithic industry of Kota Tampan is contained in a high-level river gravel ... in all probability laid down during a phase of high sea-level and the height of the terrace, 230 feet (about 70 meters) m.s.l. suggests that this high level was that of the First Interglacial". (1962: 120).

To the casual visitor these high terraces, containing the gravels and partly capped by the ash, do form the strong feature of an otherwise somewhat unimpressive near and middle distance landscape, relieved further back by the mountains bounding the valley far to the east (going up to over 2,000 meters) and the somewhat lesser range behind to the west. Some striking limestone hills (full of caves) show nearer, roughly running parallel to the Lenggong road and the river. The terrace level with the old bungalow, desolate now of human life, still commands the river and ford below. The place could well serve as a fishing camp and contact rendezvous for nomadic food-gatherers at any time. In time of strife, it would make a good look-out, and be difficult to take by surprise. As a more permanent centre Kota Tampan, in the days before good housing, cannot have been particularly attractive. The river flats below and the quickly drying, well drained surface levels above, radiate the heat quite savagely. It is not a very pleasant spot to be out-of-doors on any normally warm Malayan day.

The question then becomes: What was it really like long ago? When did this landscape, these tool-providing horizons, become accessible to early man? With no other support outside an arguable classification of tool-types (cf.Sections 6 and 7), everything hinges on geological answers.

5 RELEVANT NEW GEOLOGICAL INFORMATION

Since Neville S. Haile became, in 1967, Professor of Geology at the University of Malaya there has been a noteable stepping up of interest in and publication about later geological periods in the Malay Peninsula. Fortunately — for the present discussion — Dr Haile has interested himself particularly in Quaternary shorelines and summarised his findings in a short but essen-

tial paper (1971); while Dr P.H. Stauffer of his department has special-
ised in volcanic ash (1971, 1973).

5.1   Information on Pleistocene sea levels

Haile surveys the whole field of previous study. He emphasises in his
opening paragraph that: "there is no acceptable evidence in West Malaysia
of a sea-level stand relatively higher than the present since the Triassic"
(1971: 333). He adds that evidence put forward by previous investigators
— and hitherto generally accepted — does not stand up to critical examination. He then proceeds to that examination. He shows how the general
literature perpetuated earlier errors of sea-level attribution after these
have been corrected locally (p. 334). He considers Walker's ideas on the
old Perak 70 meters level and finds them unacceptable. And so on, until,
further: "The present writer therefore concludes, from an examination of
the literature, and from his own field-work in Selangor and Perak, that
there is no evidence for Cenozoic marine deposits in West Malaysia at an
elevation greater than 15 meters" (Haile, 1971: 335).

There is not one record, for instance, of a marine fossil from the
alluvium in Malaysian open-cast mines. Dr Haile's interesting discussion
of shallow submergence need not concern us here, since Kota Tampan as
a prehistoric site cannot be directly involved. Suffice it to cite his Conclusion (p. 341) that "there is no evidence of Quaternary submergence"
anywhere in the Peninsula beyond 6 meters.

Under these circumstances the basic Sieveking argument has to be
abandoned as regards the gravel source. Kota Tampan apparently must
have developed locally by land (not sea) level changes, erosion and other
influences, which cannot be presently related archaeologically to the First
or Second Glacials.

5.2   Information regarding volcanic ash

Even with these negative gravel results, if the overlying volcanic ash were
"old" (anytime near Middle Pleistocene) this would assist a view that the
tools below were older still. Originally, the presence of this dense ash,
usually attributed to the great Mt Toba eruption in Sumatra, was vaguely
placed in "remote prehistoric times" (Tweedie, 1953: 10).

The rhyolite-ash contains sponge-spicules, originally thought to be
marine by its discoverer in 1930, J.B. Scrivenor, who later showed, however, that it was of fresh-water origin (Scrivenor, 1949: 107; cf.Walker,
1962: 105; Haile, 1971: 334; Gobbett, 1968).

Most accounts were for Perak, until in 1971 Dr Stauffer located similar
rhyolite-ash in, for the first time, a clear stratigraphic sequence at a tin
mine at Ampang, close to Kuala Lumpur, Selangor, about 40 meters above
sea level (Stauffer, 1971: 7; cf.1973). The finder relates it to the similar
ash in Perak and elsewhere. The whole material lay between preserved
wood remains, above and below, enabling a good set of radio-carbon dates.

58

Table 1.   Summary of C-14 dates, from Ampong, K.L. in Stauffer, 1973:2

Position of wood sampled in relation to Ash layer	C-14 age (years before present)
above top	$1,145 \pm 90$
below base	$33,250 \pm 1,900$
below base	$36,500 \pm 2,500$

The indisputable logic of these results is discussed in Dr Stauffer's papers.
He favours Mt Toba; but it is unproven. Other sources are quite possible.
A potassium-argon dating result from Toba deposits in Sumatra gave
$70,000 \pm 12,000$ years, but is thought to be less reliable and quite probably
too old, since such a "young" date is not really suitable for the very long-
span potassium-argon method, as compared with short-span C-14 (Stauffer, 1973: 3). In any case, none of these dates are anywhere near Mrs
Sieveking's theoretical reconstruction of Tampanian.

The Ampang ash itself is thought by Stauffer to date close to 34,000
years. This is, of course, late Pleistocene, far younger than the Second
Glaciation. It is within the time-range of the deepest levels at the Niah
Caves in East Malaysia, which we shall look at in a moment from that
angle. It must also be added, however, that there is, as yet, no proof that
all ash fell from one and the same incident at the same time in this region.
Our Niah excavations have given contrary indications. Moreover, an
appreciable ash shower was recorded in Sarawak when Krakatau blew up
in 1883. A Sarawak museum party excavating in the northern caves wit-
nessed a volcanic ash shower on the Bakong river, 20th August 1966, and
that same day ash fell widely in the open between Miri and Niah, "causing
some alarm" (Milne, 1966: 4).

There is also an important C-14 date from charred wood underneath a
lava flow near Tawau, Sabah (East Malaysia) which gives a maximum
age of $27,000 \pm 500$ years for the lava itself (T. & B. Harrisson, 1971: 8).
No active volcanoes are known in Borneo during historic times. Thus the
situation for Malaysia (West and East) is of extensive minor volcanic ash
impact within the Last Glaciation and Post Glacial and into the present.
All this was within the span of Homo sapiens and need not raise any involve-
ment with Pithecanthropus or other hominid forms not yet proven to exist
in the region outside Java. Clearly these did occur more widely. But as
we have not a single provenly correlated artefact from Java (Harrisson,
1974) and as the Tampan tools do not clearly resemble Javan in any case,
it is inadvisable to allow these to be used as criteria for dating a "culture"
or "industry" which has now lost its previous supposed geological supports.

6 TAMPAN TOOLS IN THE LIGHT OF SOME RECENT EXCAVATIONS

The geological support for the Tampanian as an early stone-age industry is so crucial that when I wrote on this to Michael Tweedie, now retired from government service and living an active naturalist's life in England, he replied (1974): "the crucial evidence is surely that of Walker ... Walker may be wrong, of course, but a tremendous lot of measuring and surveying would be needed to show that his figures are incorrect".

That "tremendous lot" has been done by Prof Haile and his colleagues. We are left with the gravel beds intact, but no longer explicable as directly part of Pleistocene changes in sea-levels due to expansion and retraction of the polar ice-caps.

We are left, also and abidingly, with the stone tools, perhaps less to be taken for granted now that the crudeness admitted by all (including both excavators) is not automatically explained — or at least implicitly excused — by any certainty of great antiquity. It is necessary to re-examine these tools, with this position in mind, before attempting to re-assess them against a background of more recent knowledge, acquired in the Southeast Asian field since the last Kota Tampan "dig".

First a summary of Mrs Sieveking's very full presentation of her 1956 work, which also includes consideration of the earlier exercise by Dennis Collings.

Table 2.   Re-analysis of Sieveking, 1962: 113-119
Proportion of Tampan tools recovered and classified, then listed/illustrated

Her classification:

a. "More acceptable artefacts" (p. 113)	No. collected & classified	No. coded, listed & illustrated	Percentage not otherwise referred to in text
1 Points	105	27	75%
2 Cleavers	41	15	59%
3 Scrapers	36	12	66%
4 Pebble picks	12	4	66%
5 Hammerstones	1	1	–
Total "acceptable"	195	59	69%
b. "Undistinguished pieces"			
6 Undifferentiated flakes	36	1	97%
7 Rough cleavers or Heavy scrapers	23	1	96%
Total "undistinguished"	59	2	96%
Total 1 - 7:	254	61	76%

Mrs Sieveking herself describes her 96% of unpublished "rough cleavers"
as "borderline artefacts ... often made up of tabloid lumps or split pebbles"
(p.117). If these are borderline, where are the 97% unspecified "undifferentiated flakes" beyond it? So, indeed she gently implies: "These flakes
are frequently simply a slice of pebble, or a broken piece of such ...
some pieces may very well have been produced by natural causes, and
are utilizable rather than utilized". (Sieveking, 1962: 119).

The only one illustrated from this latter group (Plate 61) and there
described as a "hand axe" resembles none one has ever dared to so cat-
egorise, though she discusses this theme in some detail (nine lines of
text, p.119).

Ann Sieveking earlier and very honestly referred to the aforesaid possibility of "natural causes". She noted that "most of the artefacts from the
gravel were rolled" (p.110), an impression fully confirmed by examination.
She goes on to stress that in dealing with so poor and so primitive an
industry ... it really is not possible to distinguish between broken pebbles
and flakes that may have been utilized. She adds (frankly enough) that
"only the best of the artefacts are really acceptable as such" depending
on "context rather than appearance". That context is now no longer acceptable in the way then thought.

In his accompanying geological report, Dr Walker drew attention to the
same point when he wrote that some Tampan pebbles "are very friable,
breaking easily between the fingers" (1962: 105). The same was true in
1974. We will return to this point in final discussion (at Section 8 below).

It must also be noted that despite selection and high quality illustration,
the "hand axe" of Plate 60 is not alone in looking negative among the
chosen. It would take a true enthusiast to see a "pebble pick" in Plate 4,
thought Nos 1 and 3 are fair enough, comparable to some "chopper tools"
excavated in quite late horizons at Niah Caves (see Table 3 below). Sieveking finds these, however, "strongly evocative of pebble tools from the
early stages of the Oldowan in East Africa" (p.113), a theme she devel-
ops extensively later, finding "a remarkable similarity" with Tampan (p.
128). Plate 47, "rough cleaver", can be matched by broken and river worn
pebbles all over the place (in my view). The "cleavers" in Nos 38–40 are
very like my surface pick-ups; a multitude more available beside the road,
river rapids and elsewhere. Some of the "points on flakes", such as her
No 26, are similarly multi-repeatable?

None of this is to suggest that much from Tampan may not be human
artefact. The difficult questions remaining revolve largely round the
degree of deliberation in fracture and shaping, and whether the results
justify treatment as any sort of distinctive "culture" or "industry". If yes,
who did it, and when?

Let us for the present, look at the Tampanian another way, leaving aside
for the moment the standard 1938–56 approach, not so much because that
cannot stand re-examination as because the classical tradition of western
comparison (Abbevillian, Acheulean, Clactonian etc of Sieveking: 120–

133) grows a bit outdated in the setting of later field-work and laboratory research in Southeast Asia. In these nearly twenty years we have learned much on stone-age typology and such like. Most of what we have learned shows us how much remains to be learned; and how dangerous it can be to rush at far-reaching conclusions on inadequate supporting data. We have learned, above all, our great ignorance. We have to take a fresh look at almost everything, taking nothing from the past for granted in this region simply because it seems to work that way in another. There is nothing parochial or negative about this view. On the contrary, it requires a wider internationalism of thought, inter-change of disciplines; positivism in planning, then executing research projects along new, more sustained lines.

The Montreal Conference late in 1973 (mentioned at Section 4 above) brought out this change of mood and conviction from Pakistan and India through to Korea, Japan and the Philippines (Ikawa-Smith, 1974). Unfortunately Mrs Sieveking, though invited, was unable to attend the Montreal Conference. No one reported on or for West Malaysia.

Even the long taken-for-granted "Pithecanthropus" set-up in Java came in for considerable critical questioning (as regards real antiquity).

Four things emerged at Montreal, with near or full consensus, which relate directly to the present discussion:

1. Nowhere in this region have human artefacts yet been directly correlated with human skeletal remains and other human cultural materials in such a way as to give clear datings older than c. 30-40,000 years before present (see Niah and Tabon Caves below). Even the Java dates are, so far, derived from other, indirectly linked materials, like rock deposits, tektites, pollens, capable of subsequent (post-deposition) displacement or shift in relation to the artefacts especially (cf.Harrisson 1974 for examples of 300,000 years tektites in 40,000- year stratifications now).

2. Sites and "industries" long regarded (by tool types alone) as Middle Pleistocene or earlier have not been confirmed on other criteria, so far.

3. There is growing evidence that certain tool types continued over enormous time spans in this part of the world; and that seemingly "early" forms (choppers, flakes, etc) may continue into prehistorically quite recent situations.

4. "Western", European and American (and African) classifications, based on evolutionary concepts proven there, are seldom suitable in this region; sometimes most unsuitable, misleading and misdirecting the prehistorian.

More specifically, we now have stone-age caves in Palawan (Tabon), Sarawak (Niah), Sabah (Agop Atas) with dates going back into the Palaeolithic, the former two between 30 and 40,000 years in a time sequence, some artefacts demonstrably continuing unchanged from full Palaeolithic into early Neolithic or even later. Robert Fox, leader of the Tabon Cave excavations, has phrased it well: "Utilized flakes do not have distinct and recurring forms ... The technique of manufacture by the Palaeolithic inhabitants of Tabon Cave produced highly variable forms and sizes of flake tools. The early attempts by the writer to employ the usual typological classification and description were singularly frustrating and finally abandoned" (Fox 1970: 33).

The present writer has sought to sum up for a comparison of Palawan's Tabon (Fox, 1970) and closely linked Sabah in Northern Borneo (T. & B. Harrisson, 1971: 149-178, Figs. 14-19 and Plates 35-39), as follows:

1. There are no "prepared cores" of the kind common in many Palaeolithic areas.

2. Stones are worked in a less orderly way than usual.

3. Thus, recurring shapes and a definite typology of tools are absent, function can seldom be safely deduced from appearance alone.

4. The stones may be almost any shape permitted by the stones cleavage and other qualities.

5. Any sufficiently large surface may be used as a striking platform; and any piece of stone made into one or more tools.

6. Consequently, the range of unclear forms is great.

7. There is no long series of tools and often not even a visual similarity within a group at any one time-space location.

8. There is little or no visible change in tool techniques over millenia. (Harrisson, 1974: up-dated).

Most of these eight considerations could, in varying degrees, be applied either to part or all of the Tampan assemblage.

In the particular case of Niah Caves we have over the years excavated, among a wide range of stone tools, including many quartzite flakes (cf. 7.3 below), 18 large monofacial pebble "choppers" of the general sort first given clear focus by Dr Movius (1948) and much considered by Mrs Sieveking (1962: 112, 117, 132). These are significant here because their size and distinctive form makes them easily distinguished from anything else in the "dig". They can thus be taken as a single example to illustrate the preceding theme: that is, the wide diffusion in time of "types" often previously considered essentially "early" — and often, in fact, when found alone, treated as if much earlier than would be possible at Niah (the first place they were excavated seriously in situ with human and date associations).

Table 3.   Broad time-depth distribution of Monofacial pebble choppers in Niah Cave excavations, Sarawak (1954-67)

Approximate age of level (in years before present)	No. of tools at this approx. age level
40,000 - 20,000	9
20,000 - 5,000	6
Less than 5,000	3

Although there are less of these choppers later, with the neolithic, there are by then many more tool-types available as alternatives. And in the later part of the site, there are as many such tools as in the earlier. The essential, if not astonishing, fact is that almost identical tools of the same

stone, size, shape, are found in user-horizons from fully Palaeolithic into
the Neolithic. It is clear that under these conditions any thought of dating
a site by such — or any "primitive" — tools on their own is liable to lead
to erroneous conclusions. Stone artefacts which look very ancient (by
other standards) may not actually be so — in Southeast Asia, that is. This
is even more marked on the islands than adjacent mainland; but only as a
matter of degree. The caution cannot be ignored at Tampan in Malaya any
more than in Trinil, Java. Fresh work in West Malaysia will clarify further. There is a need for such work.

7 TAMPAN TECHNIQUES IN A WIDER CONTEXT

7.1 Bifacial and monofacial

"The most typical implements are made striking flakes off one side of the
end of a pebble". Thus Michael Tweedie (1953: 9), following his colleague
Collings in describing the Tampan material for years under his care at
Singapore. Ann Sieveking refined and developed from this basic emphasis,
but still insisted on monofacial importance, especially with the class of
"cleavers", her second largest (Table 2 above). These she equates to
Movius' earlier "choppers", "the definitive tool type of the South and East
Asian industries" (1962: 117). In this crucial group, only one (No 42) is
bifacially trimmed in the full sense.

This emphasis on monofaciality for "chopperish" pebble-tools is characteristic throughout the region. All the Niah Choppers discussed (Table 3)
are monofacial. In any further work-out of this whole problem this aspect
must be borne well in mind. The characteristic certainly does not imply
Middle Pleistocene or other early manufacture, on its own. The monofacial tendency continued powerfully even into the extensive technology of
the region's Neolithic, and is equally characteristic of the mainland
Hoabinhian (see below).

7.2 Hoabinhian — and pre-Hoabinhian?

The definition and timing of the Hoabinhian (named for 1927 cave sites
near Hanoi) and its precise relationship to the Mesolithic — "between"
Palaeolithic and Neolithic — have been much argued in recent years. I
will not enlarge on any of that here; but this difficult phase has to be mentioned, since (it seems to me) that part of the Tampan explanation may
link to the Malaysian version of this Hoabinhian.

The local Hoabinhian was earlier defined by Tweedie thus: "The hall-
mark of this culture is a type of stone implement made by flaking or chip-
ping an oval river pebble" (1953: 12).

The attentive reader will notice that if you omit "one side" from the
same author's description of the Tampanian (cited above) the result is
much the same verbally. Both "cultures" stress the use of pebbles, flaked
(or chipped).

In a thorough discussion based mainly on his field experience in Thailand, Chester Gorman listed 6 criteria for the Hoabinhian, of which the first was: "a generally unifacial flaked tool tradition made primarily on water rounded pebbles and large flakes detached from these pebbles" (Gorman, 1962: 82).

He also emphasises the use of core tools ("Sumatraliths") made by complete flaking on one side of a pebble and a high incidence of utilized flakes (cf.Tampan).

The Sumatralith is so named for its frequent occurence on that island. No clearly definable Hoabinhianism goes further East and North through the islands to Borneo and Palawan, so that nothing of the sort is present at Niah and Tabon (Harrisson, 1970). They were first recognised in Malaya from Gua Kerbau (1928) and Guah Badak (Lenggong). These cave sites are immediately along the limestone hill belt behind the Tampan stretch of the Perak River valley (cf.Section 4). As Tweedie noticed for Sumatraliths in the peninsula: "They were nearly always made from rounded pebbles of quartzite. They have been encountered only at sites in Perak, and appear to be absent from cave deposits in Kedah and Perlis and on the east side of the mountain range" (Tweedie, 1953: 12).

There is another important variant from the normal Hoabinhian (Tweedie's words) often called the "protoneolith" in Malayan literature. It is based on the handaxe, or occasionally on an unworked pebble, one end of which has been ground on both sides to give an edge. These were excavated at Gua Kerbau Perak, to the deepest levels (a good example is on display in the Muzium Negara, Kuala Lumpur) but have been found quite widely in other caves also. Other Hoabinhian types, however, have been found "at practically every cave excavation conducted in the country and also at a few open sites" (Tweedie: 12).

The Hoabinhian as a whole and as presently still uncertainly defined is largely known from caves. In Malaya it seems to cover roughly 11,000 to 5,000 years ago (cf.F.L. Dunn's very interesting, rather theoretical 1970 discussion), but evidence before 11,000 is extremely weak for this part of Asia. There is a curious, serious gap in our knowledge for the Late Pleistocene, say 150,000 to 50,000 years before present. On the mainland this continues right through up to 10,000. On the islands Tabon and Niah have helped fill in well back from 10,000, to at least c. 30,000 years — which is their importance for the region. No such sites are known in Malaya — or, for that matter, on the whole Indian sub-continent (Kennedy, 1973: 8).

It is possible that in the western and northern part of this region, the Hoabinhian should be seen not as something which evolved suddenly (and for no ascertained reason) in the Late Palaeolithic, but rather as the continuation of an already existing previous venerable tradition — in line with the long continuity of techniques, the extreme conservatism of form which has often been emphasised. In this, it is perhaps unfortunate that the Hoabinhian as such is named for and intellectually derived by the chance of early fieldwork (Madame Colani in 1927) to the north of Vietnam, under

China. It would be wrong, for that reason, to assume (as can easily
occur) that it in any way began there in earlier time. The origins may well
be further south — or anywhere. We simply have no evidence on such mat-
ters yet.

7.3 Patjitanian

At this stage, we must return to the Tampan tools themselves. To the
present writer, who has worked mainly on the eastern islands with no
visible Hoabinhian (in the accepted sense) these tools — and semi-tools
— recall Hoabinhian forms more than anything else. But some do resemble
the Javanese Patjitanian, especially in the view of Michael Tweedie (personal communication, 1974) who with Ralph von Koenigswald "discovered"
this assemblage of massive, crudely worked stone tools in Java in 1935.
These came mainly from the bed of the Baksoka River, and have regularly
been regarded as "Lower Palaeolithic" (Heekeren, 1972: 35).

However, as was demonstrated at the 1973 Montreal Conference, some
of these Baksoka artefacts are very close to those excavated in Niah Caves,
which cannot be older than Late Palaeolithic, at 40,000-years (Harrisson,
1974). A typical large Baksoka flake is exactly matched by one excavated
26.2.58 from 66"-72" (there about 25,000 B.C.) at Niah Caves (Heekeren,
1972: 37, Fig.12).

Over 95% of the Patjitanian tools are unifacial, made on or from pebbles
and flakes. Heekeren (1972: 31) considers that it differs from the earlier
tool complex of the Far East, and equates it specifically to the Tampanian.
But in Java, as in Perak, this particular "industry" remains without human
or other cultural direct correlations. In the Javanese case, however, all
are made from silicified limestone, or tuff, not quartzite. (This is
largely a matter of geological availability as well as local tradition; cf
Section 8 below). As Heekeren sums up: "It seems clear that the Patjitanian
shows evidence of a very slow tempo of change, in tool technique and
creation of new forms" (Heekeren, 1972: 43).

In view of Tweedie's great experience in Malaya, Java and elsewhere
as well as his origins as a fully qualified geologist, his views, alongside
Heekeren's deserve full consideration. These two have a wider and larger
knowledge of Southeast Asia artefacts in stone than anyone else living.

8 CONCLUSIONS: THIS "TAMPANIAN" IS WHAT? "PROTO-HOABINHIAN"?

1. The Perak River terracing can no longer be claimed, on geological
evidence, to represent First or Second Glacial (or any other) marine
deposits. Sea levels during the relevant period never reached anywhere
near the Kota Tampan height. The previous basic premise for "dating"
the Tampanian tools is therefore voided. The beds must be explained
quite otherwise, afresh.

2. The volcanic ash, by contrast, now takes on more significance, since it has provisionally been dated (by Ampong parallel) at close to 34,000 years but that could be a different "shower" (see 5.2). This makes it initially unlikely (by no means impossible) that the stone tools are much later than 35,000 years. The volcanic event it shadows might even have led to the end of the site and loss of the tool supply beds until exposed again by later erosion.

3. The form and supposed "function" of the tools themselves cannot any longer safely be taken as criteria for dating in the light of recent work on persistence of style and conservatism for change in the area (6). We have to re-think the old, classical, temperate, western-derived evolution-ary technologies back into more areal, ecological, tropical (and, inland especially, rain-forest and equatorial, pluvial) terms, somewhere along the lines of Dunn (1970) but going further yet.

4. The quality of local pebbles does not of itself facilitate refined flak-ing or close working. Allowing for that, Tampan may presently be given the benefit of the doubt and regarded, quite provisionally, as an early, crude form or pre-cursor of Hoabinhian, overlapping in time the Niahian material from East Malaysia and the end (at least) of the Patjitanian in Java (Sections 7.2 and 7.3), though for the latter exact dating remains as yet uncertain too. (This concept enlarges on one put forward for Thailand by Chester Gorman, 1972: 103).

5. Parallels between the Javanese, and specifically the Baksoka Valley finds, especially deserve further study.

6. The term "Proto-Hoabinhian" might perhaps serve until the situation is less obscure at Tampan and elsewhere.

7. This term (Proto-Hoabinhian) may — but need not — be taken to imply that the (local) Hoabinhian in Perak grew out of it, along with a dif-ferent, "more sophisticated" way of living, including especially burial in caves after c. 10,000 years before present.

8. This might also be extended to explaining, as a working theory, other open-site "stray" finds of unassociated artefacts in the region, which are not apparently "straight Hoabinhian" but nevertheless involve the frequent use of largish, monofacial tools flaked off river pebbles.

9. It would be alternatively possible, however, to look upon Tampan as perhaps an outdoor extension of the cave activity, with these gravels as workshop or alternative sources of tool supply, a little away from the easy shelter of the caves at Lenggong and all around. (Statistical re-analysis of all pre-neolithic Perak artefacts could help here?)

10. In any case, it appears more realistic, at this stage, to seek for continuities in Southeast Asia, where clearly change was exceedingly slow until much later. The theoretical elevation of single, inadequately documen-ted sites is liable to confuse eventual understanding, delay good analysis.

11. Finally, for Kota Tampan the possibilities (to put it mildly) of some "natural breakage" cannot be ignored. The pebbles there are frequently easy to break. There is also evidence that pebbles — and perhaps tools too — have been carried down from 12 miles upstream (Section 2; cf.Walker, 1962: 106).

12. This gravel formation may well be in fact due to massive erosion and transportation under high pluvial conditions. If this occurred in deep beds, underwater, in a powerful stream like the Perak River, massive pebble attrition is probable. A significant part of the more "difficult" Tampan forms might be so explained. (Experimental tests should settle this).

9 ACKNOWLEDGMENTS AND INTENTIONS

My old friends Brian Peacock, lecturer in archaeology at the Department of History, University of Malaya and Neville Haile, Professor of Geology there, have both provided valuable information, references and papers directly relevant to Kota Tampan, as discussed in the main text. Michael Tweedie, doyen of Malayan archaeologists and sometime director of the Raffles (now National) Museum, Singapore, kindly gave me his ideas on Tampan up to date and criticised my views, as part in a continuing process of interchange we have amicably conducted in various lands over thirty years. Some of the background ideas were discussed with Robert Fox of Manila and W.G. Solheim of the University of Hawaii on a recent visit to Brussels; both have wide and valued archaeological experience in the area. My colleague (since 1967) at Cornell University, Kenneth A.R. Kennedy, has given valuable clues for South Asia especially.

As indicated in the text (at Section 3) Dr F. Ikawa-Smith's conference at McGill University, Montreal in August 1973 stimulated this Tampan re-study.

The approach was extended and further defined at the Groningen symposium, 1974. The problems of volcanic ash were especially discussed with Professor H. Th. Verstappen and Dr John Flenley, though only touched on in the above text. At Groningen, too, Dr D.A. Hooijer once more put me in his debt with fresh comment on the Kinta Valley fossils, for which I have now added a separate appendix.

Raden Soejono, chief archaeologist in Indonesia, particularly emphasised (at Groningen) his personal scepticism about the "early" Tampanian in general, after a recent re-examination of the items now at Kuala Lumpur.

All through, this study has been stimulated by the parallel Indonesian activities and interests of Drs G.J. Bartstra of the University of Groningen, with whom "early" datings in the region have been closely discussed both there and at Montreal.

Meanwhile and more immediately, at and around Kota Tampan itself there is need for more research along four lines:

1. Extended search of the valley for related sites (especially upriver).
2. Critical re-examination and comparisons of the gravel and ash.
3. Statistical re-analysis of all pre-neolithic artefacts from Perak, including caves (cf.Section 8.7, above); plus some systematic wider comparisons (eg. with Patjitanian).
4. Experimental study of "natural" breakage patterns for Tampan quartz and quartzite pebbles, especially under heavy and sustained flood-water frictions (cf.Section 8.10).

68

10 BIBLIOGRAPHY

Collings, H.D., 1938. A Pleistocene site in the Malay Peninsula. Nature, 142: 575–576.

Dunn, F.L., 1970. Culture evolution in late Pleistocene and Holocene Southeast Asia. Am. Anthrop., 72: 1041–1054.

Fox, Robert, 1970. The Tabon Caves. Manila, National Museum.

Gobbett, D.J., 1968. Bibliography and index of the geology of West Malaysia and Singapore. Bull. geol. Soc. Malaysia, 2: 1–152.

Gorman, Chester F., 1970. Excavations of Spirit Cave, Thailand. Asian Perspectives 13, pp 79–107.

Haile, Neville S., 1971. Quaternary shorelines in West Malaysia. Quaternaria (Rome) 15, pp 333–343.

Harrisson, Tom, 1970. The Prehistory of Borneo. Asian Perspectives, 13: 17–45.

Harrisson, Tom, 1974. Present status and problems for Palaeolithic studies in Borneo and elsewhere. In Ikawa-Smith, 1974.

Harrisson, Tom and Barbara, 1971. The Prehistory of Sabah. Kota Kinbalu, Sabah Soc.

Heekeren, H.R. van, 1972. The Stone Age of Indonesia. The Hague, Nijhoff.

Ikawa-Smith, F., 1974. Papers of the Montreal Conference, 1973. The Hague, Mouton (in press).

Kennedy, K.A.R., 1973. Biological Anthropology of Prehistoric South Asians. Anthrop., New Delhi, 17: 1–12.

Milne, John, 1966. Earthquakes and related phenomena in North and West Borneo. Sarawak Mus. J., 14: 1–5.

Movius, Hallam, 1948. The lower Palaeolithic cultures of Southern and Eastern Asia. Trans. Am. phil. Soc., 38: 329–420.

Mulvaney, D.S., 1969. The Prehistory of Australia. London, Thames & Hudson.

Pilbeam, David, 1970. The Evolution of Man. London, Thames & Hudson.

Scrivenor, J.B., 1974. Geological and geographical evidences for changes in sea-level. J.M.B.R.A.S., 22: 107–115.

Sieveking, Ann, 1960. The Palaeolithic history of Kota Tampan, Perak. Asian Perspectives, 2: 91–102.

Sieveking, Ann, 1962. The Palaeolithic history of Kota Tampan, Perak. Proc. Prehistoric Soc., 28: 103–139. (cf.Walker).

Stauffer, P.H., 1971. Quaternary volcanic ash at Ampang. Geol. Soc. Malaysia, Newsletter 23: 5–8.

Stauffer, P.H., 1973. Late Pleistocene age for volcanic ash in West Malaysia. Geol. Soc. Malaysia, Newsletter 40: 1–4.

Tate, R.B., 1970. Tektites in Brunei. Brunei Mus. J., 2: 253–259.

Tweedie, M.W.F., 1953. The Stone Age in Malaya. J.M.B.R.A.S., (Special Monograph) 26: 1–90.

Walker, D.C., 1956. Studies in the Quaternary of the Malayan Peninsula — I. Federation Mus. J., 1–2: 19–34.

Walker, D.C., 1962. The Palaeolithic industry of Kota Tampan, 1. In Sieveking, 1962.

APPENDIX: PERAK FOSSILS AND TAMPAN

My old colleague Lord Medway has listed evidence for nine localised extinct faunas so far known in Southeast — 5 in Java, 1 in Borneo (Niah), 2 in Celebes and 1 in Malaya. This last he terms "Kinta Valley", from the Perak River tributary of that name (his No 4 at p 79 of Quaternary and Recent Mammal Faunas of Malasia, an important contribution to Transactions of the Second Aberdeen–Hull Symposia on Malesia Ecology, Hull (Dept of Geography; Misc.Series No 13, 1972)).

He regards the Kinta Valley fauna as Middle Pleistocene, with two main source sites: Tambun Cave, near Ipoh and Lake Chenderoh, south of Tampan.

From Lake Chenderoh comes one tooth of a large Probiscidean, regarded as Palaeloxodon namadicus, found casually in a tin-mine at Salak, to the south of the lake. Recent geological research, already cited in the main text, has not any clear dating for these mine deposits, as Professor Haile confirms. In any case there is no geological connection, apparently, between these and the Tampan gravels.

F.L. Dunn and Brian Peacock also cast doubt on the relevance and identity of this tooth on other grounds (J.M.B.R.A.S. 41, 1968: 171–179), though Medway has not included this reference. I have further doubted the validity of dating anything from single probiscidean molars in the region, on the basis of a re-examination of evidence on the three known from Borneo (Federation Mus. J.; in press; and see Harrisson 1974 in main bibliography above). Medway, however, is inclined to accept one tooth from Brunei (Sarawak Mus. J., 20, 1971: 346).

The other 7 species were found together deep in a fissure of a limestone cave in the Kinta Valley near Ipoh. Some are "characteristically Middle Pleistocene", according to Dr D.A. Hooijer, who has described the series (Federation Mus. J., 7, 1962: 1–5), and kindly supplemented that paper by personal communication (1974).

Tambun is 51 km south-east of Tampan. It has as yet produced no archaeological evidence, nor artefacts, though other caves in the area contain (as above noted), strongly Mesolithic or very late Palaeolithic materials (Hoabinhian, etc.). The fossils include teeth and other fragments of the still extant Rhinoceros sondaicus, one tooth of an unidentified pig, two teeth which may be of the Sambhur Deer (Rusa unicolor), one vertebra of a small carnivore (perhaps Cynogale, the Civet Otter), and parts of a large bovine which could be Seladang or water buffalo.

The two surely "older" elements in this strikingly mixed small assemblage are Duboisa santeng (two teeth of this extinct ox, also known from Java) and an unidentified Hippopotamus (shaft bone). Only these two are certainly extinct in Malaysia, though it is always possible they continued into the late Pleistocene there, like the Giant Pangolin (Manis palaeojavanica) of Java's Trinil which occurs in non-fossil form later than 40,000 years BP at Niah Cave (Harrisson, 1974 for the implications in dating Southeast Asian fossil beds by association from elsewhere).

There is no presently demonstrated relationship, date-wise, between these limestone caves and the Tampan deposit.

Figure 3. Nepenthes ampullaria Jack

Figure 2. Type of aerial root system in peat forest

Figure 5. Layer of grey volcanic ash, intercalated between peat, Rawa Lakbok
Figure 6. Coconut on peat, Sumatra

B. POLAK
Wageningen

5

CHARACTER AND OCCURRENCE OF PEAT DEPOSITS IN THE MALAYSIAN TROPICS

1 INTRODUCTION

For a long time the opinion has prevailed that high temperatures in tropical lowlands and consequently high bacterial activity along with total evaporation of rainfall should prevent the accumulation of plant residues. That
is why it was generally accepted that peat deposits in the tropics can only
be found at higher altitudes in the mountains, where temperatures are
fairly low. However, in nineteenth century local scientific periodicals on
Java, finds of peat in the lowlands had been mentioned several times, but
apparently those articles went unnoticed outside Indonesia.

In 1895 the botanist S.H. Koorders (vide IJzerman) published the
accounts of an expedition across Sumatra. When descending from the central mountains to the eastern coastal plain, he had to traverse extensive,
very swampy forests thriving on peat. His description of the expedition
drew the attention of the palaeobotanist H. Potonié in Berlin, who together
with Koorders in 1905 published a paper about the character and composition of those peat formations.

Whereas it had been considered an impossibility for peat to form under
tropical conditions, the modified conclusion had now been arrived at, that
Sumatra peats must be of the "Lowmoor" type, deposited under stagnant
water; the preservation of organic residues under water being considered
the only possibility. From now on it became generally accepted that in
tropical lowlands peats might be found, but only in the topogenous form.
Also coal with the same restrictions was considered to be of tropical
origin; hitherto the belief of the nonexistence of recent peat in the tropics
had also denied this possibility.

2 TYPES OF PEAT IN INDONESIA

In the first half of this century, between the two world wars, the forestry
service circumstantially explored the Indonesian forests and thus it came
to light that large areas along the east coast of Sumatra, the west, north,

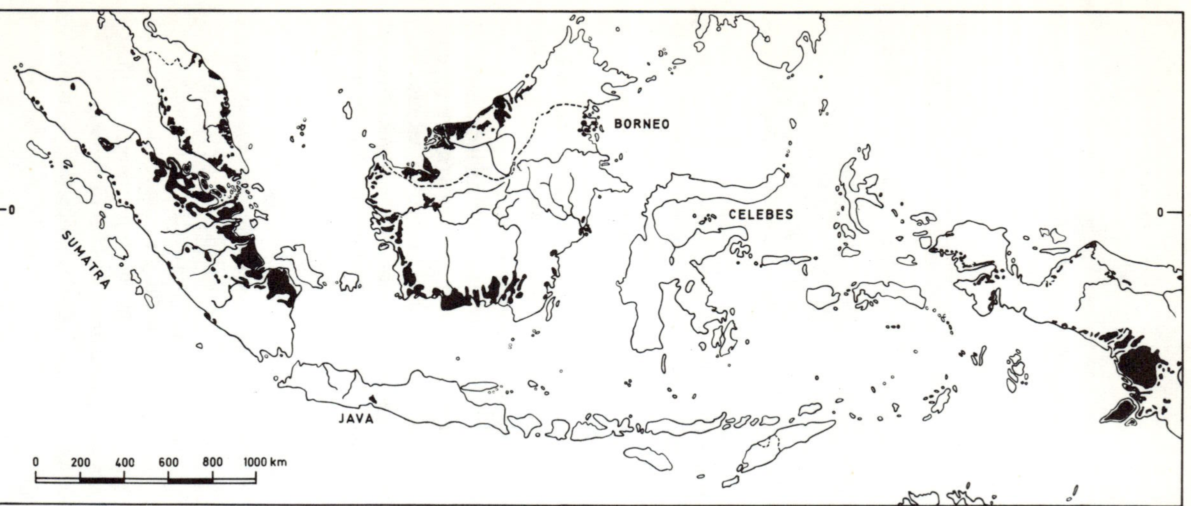

Figure 1. Peat deposits (black) in Indonesia (partly after Andriesse, 1974)

and south coasts of Borneo and both sides of the Malayan peninsula, were
covered with forests on thick peat layers. Those deposits are situated
landinward of the coastal saltwater swamps; covering an enormous ter-
ritory, roughly estimated exceeding 16 millions of hectares. Besides
those regional peat formations, smaller local peats are found in various
places. On Java only local topogenous deposits are found. In the mountains, at higher altitudes, small peat bogs are rather common in depressions of the soil. Even Sphagnum mosses are found, but curiously enough,
they grow directly on firm subsoil and do not give rise to any deposit.
The same phenomenon is mentioned by Skotsberg (1939) for Hawaï.

3 THE REGIONAL PEAT FORESTS

The extensive peat forests in the coastal plains must be considered as one
of the wonders of tropical nature, offering many a biological problem.
Those woods, thriving on their own debris, form a close biological system in which very special conditions prevail, determining the properties
of the living vegetation as well as of the organic layers produced by it.

The forest soil consists of the remains of former generations, decayed
into a brown mash and held together by a framework of tree branches and
trunks. The semi-liquid layers sometimes attaining a thickness of 15 m
or more, are composed of a high-organic substance of high acidity (pH
3–4); the ash content is about 3% . Saturated with water, the deposit is
very poor in oxygen, so that the trees develop aerial roots of different
kinds and shapes. Walking in those forests, one hardly touches the ground,
but one actually walks on a network of tree roots emerging from it. It is
the same adaptation to a poorly aerated soil as in the salt-water mangroves,
but the species belonging to it are quite different. The vegetation is composed of trees of the tropical rain forest adapted to the specific conditions
of this particular environment.

The botanical composition of the recent forests is rather well known.
The investigations done by the forestry service were carried out on behalf
of stocktaking of timber-trees for which a market was found in Singapore.
For this purpose trees with a trunk diameter of over 40 cm were collected;
ferns, herbacious angiosperms, epiphytes and undergrowth remained
unidentified. In 1938 Sewandono counted for the peat which convered the
island of Bengkalis near the east coast of Sumatra 62 tree species and 80
in the coastal plain itself.

In a detailed floristic report Anderson (1963) mentions 142 tree species
in north-west Borneo, 38 small slender species and he also identified the
herbaceous components. It may be expected that there will be a great con-
formity with Sumatra.

In 1939 Buwalda did some explorations in east Sumatra (Indragiri)
along the Kwanten river and its branches. He made extensive vegetational
records and found differences in plant communities dependent on the
thickness of the peat layer and the distance of the river. On layers of over

3 m deep, vegetation is poorer in species than on deposits of less depth.
On very thick peat deposits Myrtaceae and Calophyllum species with tall
slender trunks growing close to each other, dominate. In the innerpart of
the forest, the thickest layers show a more open vegetation with poorly
developed, twisted and stunted trees and scattered small pools containing
deep brown water. The acidity amounts to pH 3–3.5. The Myrtaceae-
Calophyllum forest is rich in Nepenthaceae, whilst mosses, ferns and
specific Cyperaceae cover the soil. On peat deposits less than 3 m deep,
the undergrowth consists of Araceae, Commelinaceae, Palmae (Zalacca
conferta, Licuala) and ferns. The pH varies from 3.5 to 4.5.

A zonal differentiation dependent on the thickness of the depsoit and its
drainage can be distinguished. For Indragiri Buwalda mentions six different vegetation types, with a dominance of one or more species. For Borneo Anderson (1963) described a similar situation.

4   CHARACTER OF THE DEPOSIT

4.1   Chemical properties

As deep as borings could be done (up to 6 m and with great difficulties,
owing to the high content of partly or non-decomposed wood), the peat
appeared to consist solely of debris of forest vegetation, one generation
piled upon the other. The peat substance is soaked with water, is low in
mineral content and is very acid; the lime content never exceeds 0.5%.

The content of plant nutrients shows much resemblance to that of
European bog peat, but is often somewhat lower. The tropical samples
are taken from spots where tidal influence is still active. Material from
the interior, where drainage is obstructed and vegetation stunted, was not
available but it would have shown still lower figures.

The composition of the organic substance is different in that it is rich
in lignin, whilst the percentages of water-soluble compounds and of hemi-
cellulose, cellulose and protein are low. Probably the high temperatures
cause a rather strong bacterial activity, resulting in the decomposition of
cellulose and hemi-cellulose, which compounds are rather stable in temperate climates. The high lignin content must be due to forest vegetation,
whilst the European highmoor peats originated from less woody plant
associations.

4.2   Botanical remains in the peat substance

As was mentioned before, the peat layers are very rich in wood, more or
less decomposed trunks and branches; seeds, fragments of fruits and lumps
of resin are also found. The amorphous semi-liquid peat substance contains
the same kind of residues as those that are found in European peats, viz:
moulds, hyphae, fungal spores and sporangia, scales of rhizopods and
arthropods; further vessels, tracheids and other fibres, rootlets and root-
tips, sometimes covered by fungal hyphae (Mycorrhiza), leaf fragments,

74

Table 1. Plant nutrients in tropical forest peat (Hardon and Polak, 1941)

Sample number	Origin	Peat type	N	Ash	K_2O	CaO	P_2O_5
			in percentages of dry matter				
94853	Sumatra	acid forest peat	1.06	3.00	0.21	0.326	0.04
94919	Borneo		1.17	1.27	0.11	0.345	0.03
94924	Borneo		0.93	0.72	0.07	0.139	0.01

Table 2. Plant nutrients in European oligotrophic peat

	N	Ash	K_2O	CaO	P_2O_5
	in percentages of dry matter				
Moss peat	0.08	2	0.03	0.25	0.05
Ericaceae peat	1.3	3	0.05	0.35	0.10

Table 3. Chemical composition of tropical forest peat, in percentages of dry matter (Hardon and Polak, 1941)

Sample number	Origin	Peat type	Soluble in:			Hemi-cellulose	Cellulose	Lignin	Protein
			ether	alcohol	water				
94853	Sumatra	acid forest peat	4.67	4.75	1.87	1.95	10.61	63.99	4.41
94919	Borneo		2.50	6.65	0.87	1.95	3.61	73.67	3.85
94924	Borneo		2.85	3.63	0.56	0.73	0.21	68.89	3.97

Table 4. Chemical composition of European bog peat in percentages of dry matter (Kivinen, 1938)

Origin	Peat type	Soluble in:			Hemi-cellulose	Cellulose	Lignin	Protein
		ether	alcohol	water				
Finland	bog peat	3.5	4.6	7.8	18.2	16.6	38.5	3.8
	forest-Sphagnum peat	2.8	5.8	3.6	12.2	4.4	38.4	9.4

cuticles and epiderms, plant hairs, fern spores and sporangia, spores of desmids and quantities of well preserved pollen grains.

4.3   Other properties

The entire deposit is dome-shaped, which can be observed in the field, when the forest is newly cleared. One sees the bare brown peaty soil rise above the adjacent mineral land. It has been proved by levelling and also by rivers and streamlets which emerge from it and flow down on all sides. These waters are of a deep brown colour and of high acidity. Locally they are known as "black rivers" (aer itam).

All those qualities together unmistakably give these deposits the character of a raised bog, i.e. one of ombrogenous origin, fed by rain water.
However the vegetation differs greatly from that known of the "highmoor"
in temperate regions. The shape, development, and content of plant nutrients are much the same. This means that the peat was deposited above
the ground-water level and that consequently the plant remains have been
preserved in atmospheric water. This is in contrast with the then prevailing opinion that in the tropics evaporation of rainfall, oxidation and bacterial activity are so intense, that peat can only be deposited if covered by
stagnant water.

However, important facts have been overlooked in formulating those
restrictions.

First of all, the ombrogenous peat deposits are found in regions where
the tropical climate is pronouncedly maritime, with a high amount of rainfall all the year round (over 3 m = 120 inches yearly). There is no marked
difference between the dry and the wet season, that is why there is no desiccation of the soil. The heavy foliage protects the peat from insolation and
thus brings about saturation of the air. Tropical vegetation, not interrup-
ted by a seasonal standstill, has a continuous production.

Microbiological activity is slowed down by a marked humidity and acidity
of the soil; that is why the destruction of organic matter, although intensified by high temperatures, is surpassed by a steady production of plant
material and so peat accumulates.

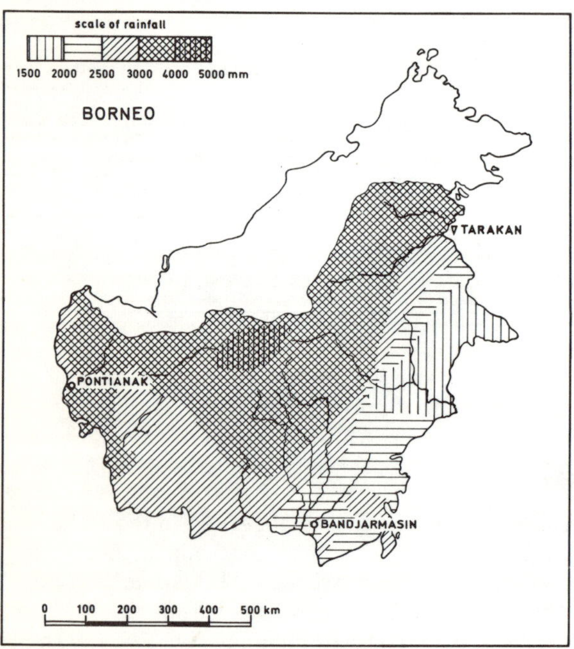

Figure 4. Borneo, mean annual rainfall

76

A striking feature is the geographical distribution. The forest peats are
found at the margins of the western islands (eastern Sumatra, Malacca,
western and southern Borneo). In the past, during Pleistocene and early
Holocene times, those areas were connected by the Sunda shelf. The rising
of the sea-level separated them. The influx of the sea caused a rise of the
ground-water level on the shores, and thus marshy or swamplike alluvial
plains developed, which formed the initial stage of the future peat deposits,
the bottom of which consists of lixiviated silts.

This development shows great resemblance to the history of the peat
deposits along the North Sea coast of the Netherlands and Flanders. The
North Sea was dry land during glacials in the Pleistocene and during the
early Holocene period. The melting of the icecaps caused the water-level
to rise, the sea came in and on the silts along the shores marsh peat
developed on a brackish reed-vegetation, which in a maritime climate developed into ombrogenous bog peat.

5 TOPOGENOUS DEPOSITS (LOWMOOR OR MARSH PEAT)

Peat layers deposited in local depressions of the soil are found on all the
islands, both in the plains and the mountains. Usually those peats are
eutrophic, but if the water of the basin in which peat originates is poor in
minerals, the deposit will be oligotrophic. Those topogenous peats show
great resemblance to lowland marshes and swamps found in various other
regions and climates; the characteristic vegetation is composed of reeds,
sedges, aquatic plants, ferns and plankton organisms, while shrubs and
trees are found as well.

This is the only type of peat found on Java. The content of minerals,
especially lime is higher than in the ombrogenous forest peats.

The Rawa Lakbok on Java, in the former Residency of Priangan, near
the south coast, was more closely investigated (Polak 1949). A basin-
like depression is filled with peat layers, which at certain spots reach a
depth of over 6 m (average depth 3 m) and cover nearly 3000 ha area
(about 7500 acres). A normal succession in the composition, according
to which the marsh vegetation gradually rises above water-level, whilst
its character and composition changes, could not be established.

In many sections layers rich in forest remains give way to peat mainly

Table 5. Plant nutrients in tropical marsh peat

Sample number	Origin	Peat type	N	Ash	K_2O	CaO	P_2O_5
			in percentages of dry matter				
61835	Rawa Lakbok Java	marsh peat	1.79	17.0	0.10	2.51	0.19
61837	"		1.60	37.4	0.13	2.34	0.36
15716	Borneo		2.46	16.5	0.11	3.36	0.11

consisting of Gramineae, Cyperaceae, Pteridophytae, and aquatic plants. This shows that, at some time in the development, the water-level must have risen.

Another peculiarity of the Rawa Lakbok peat is the presence of a layer of pure volcanic ash (average thickness 40-45 cm = about 18 inches), intersected in the peat at a depth of about 1 m (40 inches) all over the Rawa. It is highly probable that this layer upset the drainage equilibrium, thus causing the inverse succession in vegetation.

It has not been possible to determine from what volcano this ash layer originated. As radiocarbon dating had not been introduced at the time of the investigation, it could not be established when the tremenduous erpution which produced the ashes had taken place.

6 AGRICULTURE

Because of the increase of the population in the beginning of this century, there was a greater need of arable land and so the people slowly began the occupation of the margins of the peat areas of Sumatra and Borneo. They did so in a way that has been practiced on peat soils all over the world by all kinds of people: i.e. by clearing the vegetation, draining and burning the top layers.

At first annual crops were planted, such as : pine-apple, cassava, gourds, corn and beans. Because of regular burning and heavy drainage, subsidence is considerable. Insolation, irreversible drying out and oxydation of the surface aggravates shrinkage, which may result in extensive fern wastes.

During the process of subsidence often permanent crops like coconut and rubber were planted, which through shrinkage lost the soil at their roots. Especially coconut palms suffer badly from these deficient reclamations; they fall down and by means of new adventitious roots, the trunks rise again. Thus older coconut plantations offer a curious sight of coiled palms bending on all sides. The soil around the rubber trees shrinks from the roots and the high, slanting and stilted trunks hook their branches into each other for support.

The lifetime of such plantations is fifteen years at the most. In the beginning crops thrive well, but after some ten years they yellow and decay. Plantations in decline and subsequently left to become a nuisance to surrounding cultivated areas, as they are a breeding place for diseases and pests and a favourite haunt of wild animals, which ravage the neighbourhood.

On Java something similar took place. Until 1924 the 3000 ha of the Rawah Lakbok were an impenetrable jungle, consisting of swamps, marshes and open expanses of water. Reclamation meant the end of nearly all the vegetation there was; but extensive drainage was procured through the cutting of canals and ditches.

Rice and on a lesser scale sweet potatoes were the main crops under

78

cultivation. The soil is richer in plant nutrients than the Sumatra and Borneo peat and so for the first ten years crops throve, but after some years yields decreased gradually. In 1941 the situation was far from ideal. In this region, with a distinctly dry season, drainage had been achieved to such an extent, that the crops depended entirely on rainfall. In the wet season heavy damage was caused by inundation and in the dry season shortage of water became a dominant factor. Besides, the losses caused by diseases and pests were quite considerable.

Especially in the interior, on thick peat layers the rice crop looked very poor. The plants were badly developed, very low in stature with brown-spotted or yellow leaves. Most of them proved to be sterile or to have flowered at a time when normal paddy could be reaped.

Besides the problems in connection with drainage and inundation one should remember that no fertilizers had been added to the soil for over 15 years. Lack of fertilizers is also the main drawback with the acid peat of Sumatra and Borneo.

In 1948 pot and field experiments were started on acid peat from Borneo (Ehrencron, 1949, Polak and Soepraptohardjo, 1951). A small variety of maize (Madura var.) was used as a test-plant; pH of the peat was 3.3. Liming to effect an increase of pH value of at least 5.5, proved to be necessary for normal growth. The application of rather large quantities of N.P.K., together with minor elements, especially copper, ensures healthy plants with well-seeded ears.

Many difficulties were encountered in the lay-out of field trials and the data on these experiments only allow some general conclusions (Polak and Soepraptohardjo, 1951). It may be said that quantities of lime 15,000 kg/ha, ammonium sulphate 200 kg/ha, double super-phosphate 250 kg/ha, potassium sulphate 200 kg/ha and copper sulphate 50 kg/ha yielded mediocre results.

7 CONCLUSIONS

It appears that the problem of the reclamation of the extensive peat swamps on the coasts of Sumatra, Malacca and Borneo is still unsolved; this problem entails great economical and technical consequences.

A detailed research has to be conducted as to the application of fertilizer and minor elements to different types of peat in relation to various crops; costs of fertilization will be high.

Drainage is another problem; removal of trunks is expensive and difficult; too much lowering of the water-level may cause considerable subsidence and shrinkage, leading to irreversible desiccation, which renders the soil unsuitable for agriculture. The danger of flooding cannot be excluded either, whilst too dry a surface furthers the powdering of peat material and the nuisance of dust. Grand-scale techniques, causing total destruction of the natural vegetation and the underlying peat might become disastrous.

Moreover, it must be understood, that the coastal forest peats, although

they cover large areas, are a rare phenomenon in the world. Their origin
depends on the presence and co-operation of specific conditions, such as a
tropical climate with a high amount of rainfall all through the year, and a
rise of the ground-water level. Those restrictions brought about a rather
vigorous plant community, thriving on its own debris which contains a
minimum of plant-nutrients and is wetted by atmospheric water only.

Under such extreme circumstances the ecology raises many a biological question, such as the circulation of the minerals, the providing of
nitrogen by means of mycorrhiza and nitrogen-fixing microbes and the role
of the Nepenthaceae. There is also the natural selection of the species
which make up the community and their inventarisation, and last but not
least, the botanical and geological history of those formations, and their
chemical composition.

In our time, exploitation and conservation of nature are opposed to each
other. Exploitation of the extensive peat forests is difficult and expensive
and cannot be very profitable. It should be recommended to leave the
regional peat forests in their natural state or use them for timber-growing
with the application of proper silvicultural practices. These practices may
not involve measures interfering with the natural environment (Andriesse,
1974).

REFERENCES

Anderson, J.A.R., 1964. The structure and development of the peat
    swamps of Sarawak and Brunei. J. trop. Geogr., 18.
Andriesse, J.P., 1974. Tropical lowland peats in South-East Asia. Dep.
    Agr. Res. R. trop. Inst. Amst. Commun., 63.
Buwalda, P., 1940. Bosverkenning in de Indragirische Bovenlanden. Rep.
    For. Res. Sta. Bogor (not in print).
Ehrencron, V.K.R., 1949. Oriënterende potproeven met zuur bosveen van
    Borneo. Contr. gen. Agr. Res. Sta. Bogor, 80.
Fleischer, M., 1891. Die Eigenschaften des Hochmoorbodens als landwirt-
    schaftliches Kulturmedium. Landw. Jb., 20.
Hardon, H.J. and B. Polak, 1941. De chemische samenstelling van enkele
    venen in Ned. Indië. Meded. Lab. Scheik. Onderz., 101, Bogor (Land-
    bouw 17).
Kivinen, E., 1938. Ueber die Moorböden Finlands. Trans. Sec. Comm.
    Alkali Subcomm. Int. Soc. Soil Sci., Helsinki, A. 118.
Koorders, S.H., 1895. (vide IJzerman)
Polak, B., 1933. Ueber Torf und Moor in Niederländisch Indien. Verh. K.
    Akad. Wet., Amsterdam, 30.
Polak, B., 1941. Veenonderzoek in Ned. Indië: Stand en exposé der vraag-
    stukken. Contr. gen. Agr. Res. Sta. Bogor, 53.
Polak, B., 1946. Over veenonderzoek in Ned. Indië. Natuurw. Tijdschr.
    Ned. Indië, 102.
Polak, B., 1948. Landbouw op veengronden. Landbouw, 20.

Polak, B., 1948. Waarnemingen betreffende het gedrag van cultuurgewassen op veen. Landbouw, 20.

Polak, B., 1949. De Rawa Lakbok, een eutroof laagveen op Java. Contr. gen. Agr. Res. Sta. Bogor, 82.

Polak, B., 1950. Occurrence and fertility of tropical peat soils in Indonesia. Fourth int. Congr. Soil. Sci. Amst.

Polak, B. and M. Suprapotohardjo, 1951. Pot and field experiments with maize on acid forest peat from Borneo. Contr. Agr. Res. Sta. Bogor, 125.

Potonié, H. and S.A. Koorders, 1905. Die Sumpfflachmoor Natur des produktiven Carbons. Jahrb. K. Pr. Geol. Landesanstalt, Berlin.

Sewandono, M., 1938. Het veengebied van Bengkalis. Tectona, 31.

IJzerman, F.W., 1895. Dwars door Sumatra. Haarlem — Batavia.

JAN MULLER
Rijksherbarium, Leiden

6

POLLEN ANALYTICAL STUDIES OF PEAT AND COAL FROM NORTHWEST BORNEO

INTRODUCTION

The pioneer researches of Dr B. Polak (1933) first demonstrated the presence of an abundant and well preserved pollen and spore microflora in Malesian peat deposits. This opened up the possibility of stratigraphic analysis, provided an adequate reference collection of correctly identified recent plant species as well as sufficient ecological data on the plant communities responsible for peat accumulation were available. It was only after the second world war that this information could be obtained when Dr J.A.R. Anderson embarked upon a detailed survey of the coastal peat swamps of Sarawak, which cover more than 15,000 km^2. In 1958 the first boring was made near Marudi (Sarawak), Fig. 1, and the results have been incorporated in a paper by Anderson and Muller (1975). Subsequently other peat deposits were studied and, in an attempt to trace the pre-Quaternary development of peat swamp vegetation, a Miocene coal deposit was also investigated.

This note intends to summarize the results obtained.

METHODS

The most difficult aspect of the project proved to be the boring. Due to the high content of only partially decayed wood, even the heavy Kjellerman piston boring equipment had to be shifted several times before the peat deposit could be penetrated and the underlaying clay deposit be reached.

A second problem, limiting the application of pollen analytical methods, is the restricted dispersal of most of the pollen grains and spores produced by the peat swamp vegetation which mainly consists of forests. A limited amount of surface samples were studied and the results showed that their pollen content only reflected the composition of the local vegetation. To study the regional development of the peat formation, a series of borings would be necessary, the spacing and sampling density of which should cancel out the local factors. This should be based on a detailed survey of the

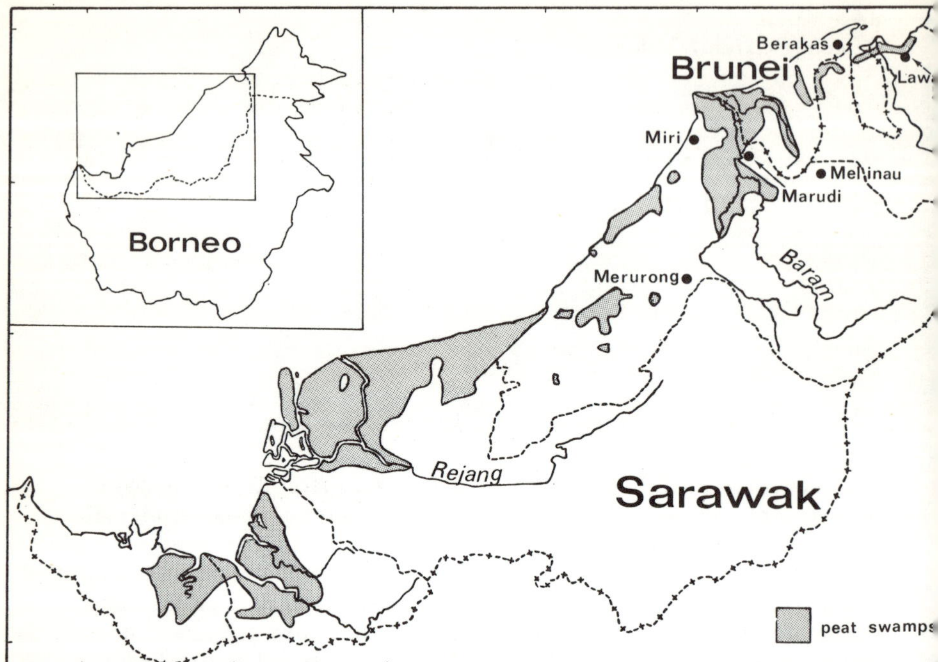

Figure 1. Location of the sample spots.

pollen content of the surface layer correlated with floristic composition at
the sample spots.

It will be clear that the results will not yield much information on the
vegetation history outside the peat swamps and will only reflect climatic
history in so far as it has influenced the succession of peat swamp communities. Since peat accumulation in the lowland tropics only takes place
under everwhet climatic conditions, a change in climate can only be towards
less humid, either average or seasonal conditions, which will tend to
decrease the probability of peat accumulation and would lead to a hiatus.

Far more likely to be present and difficult to separate from any climatic
causes are edaphic changes caused by subsidence and marine transgression or elevation and increased drainage of the swamp surface. Such
changes will manifest themselves as abnormal successions, e.g. reversion
to an earlier stage.

RESULTS OF THE MARUDI BORING

The peat deposit proved to be 11 m thick and rested on a clay, deposited
in a mangrove environment. Peat accumulation started approximately 4000
years ago and this dating suggests a connection with the termination of the

84

postglacial eustatic rise in sea-level. More borings will be required to establish the regional timing.

The pollen content of the peat indicated a succession which broadly compared with the zonation present at the surface. One significant exception was present about midway, where an abundance of Garcinia cuspidata pollen suggested a disturbance of the succession, which because of lack of a recent counterpart, could not be satisfactorily explained as yet. Future borings will have to show whether this phase is of regional extent and what could be its ecologic significance.

It is also worth noting that in the Marudi profile which is only one mile away from the Baram river, no trace of its influence was visible. This is clear evidence that the convex raised bog surface has effectively kept the river confined to a rather stable meander belt.

RESULTS FROM OTHER AREAS

A second boring was made in a small coastal peat swamp near Lawas (Sarawak) which differs from the Marudi swamp in dominance of Dacrydium and Gymnostoma (= Casuarina). The results (unpublished) showed the deposit to be approximately 3.5 m thick and to grade into the underlying sandy sediments which contained mangrove pollen; the two genera mentioned proved to be dominant throughout the peat deposit which showed few successional changes.

Quite different are the small peat deposits which have accumulated in poorly drained depressions on the sandy, podzolic soils of Kerangas forests and which were studied in collaboration with Dr E.F. Brunig. The influence of the surrounding vegetation is more clearly expressed in the pollen content than in the vast coastal peat deposits. It is therefore difficult to decide how far the succession evident in these deposits reflects the peat swamp succession or edaphic changes in the surrounding vegetation. The thickness of these Kerangas peat deposits is less than of the coastal peats. In the Sungei Dalam forest reserve near Miri (Sarawak) a 110 cm thick deposit could be investigated which contained abundant pollen of Dacrydium, Calophyllum and Sapotaceae and an increase upwards of Gymnostoma pollen. In the Melinau area (Sarawak) a 150 cm thick deposit was characterized by dominance of Calophyllum and Sapotaceae pollen, while Gymnostoma showed a decrease upwards.

On the Merurong plateau (Sarawak) three peat swamps were sampled ranging in altitude from 730 m to 1150 m, with a thickness varying from 25-100 cm. Characteristic pollen types were Gymnostoma, Dacrydium, Eugcissona insignis, Melanorrhoea, Ericaceae, and Garcinia cuspidata.

ORIGIN OF THE PEAT SWAMP FLORA

It has already been emphasized by Anderson that peat swamp endemics do not exist and that nearly all species occurring in peat swamp forests also occur in Kerangas forests, the podsolized soils of which are as poor in nutrients as the peat swamp soils. It thus appears likely that the type of vegetation covering the newly developing peat areas is initially dependent on the composition of the nearby Kerangas forests which act as reservoirs. This is not surprising in view of the, geologically speaking, ephemeral nature of the peat swamp environment, dependent as it is on the coincidence of a number of physiographic and climatic factors.

The presence of coal deposits in the Tertiary sediments of the NW Bornean geosyncline provides a unique opportunity to compare the composition of peat swamp communities of earlier times with those of the present. As a first attempt, a coal deposit near Berakas (Brunei) of Miocene age has been investigated. As reported in Anderson and Muller (1975), the composition of the swamp flora of this deposit proved to be closely comparable to the initial mixed swamp forest phase of the recent peat swamp zonation. The abundance of Melanorrhoea and Cephalomappa, two genera which are conspicuously absent in the Marudi swamp make the Berakas deposit comparable to the recent Rejang swamps, while the abundance of Durio finds its counterpart in some peat swamps along the coast of the Malay Peninsula. The Berakas swamp has been small in extent. It started on fluviatile sediments, but already during its development regular influx of mangrove pollen attests to the nearness of the coast. After a development of max. 10 m of peat marine transgression terminated peat growth.

CONCLUSIONS

The pollen and spore content of Malesian lowland peat and coal deposits makes it possible to arrive at a fairly accurate reconstruction of the past succession of plant communities. From the Miocene onwards the overall composition of these peat swamp communities has not changed much. Local variations may partly be attributed to accessibility of the developing peat swamp environment for plants from the nearby Kerangas forests which act as reservoirs.

REFERENCES

Anderson, J.A.R. and Jan Muller, 1975. Palynological study of a Holocene peat and a Miocene coal deposit from NW Borneo. Rev. Palaeobot. Palynol. (in print).

Polak, B., 1933. Ueber Torf und Moor in Niederländisch Indien. Verh. K. Akad. Wet. Amst., Afd. Natuurk. (2e Sectie), 30(3): 1-84.